BITS OF POWER

Issues
in Global
Access to
Scientific
Data

Committee on Issues in the Transborder Flow of Scientific Data

U.S. National Committee for CODATA

Commission on Physical Sciences, Mathematics, and Applications

National Research Council

NATIONAL ACADEMY PRESS
Washington, D.C. 1997

NATIONAL ACADEMY PRESS • 2101 Constitution Avenue, NW • Washington, DC 20418

NOTICE: The project that is the subject of this report was approved by the Governing Board of the National Research Council, whose members are drawn from the councils of the National Academy of Sciences, the National Academy of Engineering, and the Institute of Medicine. The members of the committee responsible for the report were chosen for their special competences and with regard for appropriate balance.

This report has been reviewed by a group other than the authors according to procedures approved by a Report Review Committee consisting of members of the National Academy of Sciences, the National Academy of Engineering, and the Institute of Medicine.

Support for this project was provided by the National Science Foundation (under grant no. INT-9507279), the National Library of Medicine (under purchase orders 467-FZ-501994, 467-MZ-600631, and 467-FZ-402090), the Defense Technical Information Center (under purchase order no. D-03-94 and award no. SP060096P0470), the National Aeronautics and Space Administration (under purchase order S-73093-Z), the National Institute of Standards and Technology (under contract no. 50SBNB6C9118), the National Oceanic and Atmospheric Administration (under grant nos. NA56EC0622 and 56-DKNA-7-95101), and the Department of Energy (under grant no. DE-FG02-96ER30277).

Library of Congress Cataloging-in-Publication Data

Bits of power : issues in global access to scientific data / Committee on Issues in the
   Transborder Flow of Scientific Data, U.S. National Committee for CODATA, Commission on
   Physical Sciences, Mathematics, and Applications, National Research Council.
      p. cm.
   Includes bibliographical references and index.
   ISBN 0-309-05635-7
   1. Communication in science. 2. Research—Information services. 3. Database
management. 4. Information technology. I. U.S. National Committee for CODATA.
Committee on Issues in the Transborder Flow of Scientific Data.
   Q223.B58 1997   97-4836

Bits of Power: Issues in Global Access to Scientific Data is available from the National Academy Press, 2101 Constitution Ave., NW, Box 285, Washington, DC 20055 (1-800-824-6242; http://www.nap.edu).

Cover photo courtesy of Peter Arnold, Inc., New York, N.Y.

Copyright 1997 by the National Academy of Sciences. All rights reserved.

Printed in the United States of America

COMMITTEE ON ISSUES IN THE TRANSBORDER FLOW OF SCIENTIFIC DATA

R. STEPHEN BERRY, University of Chicago, *Chair*
SHELTON A. ALEXANDER, Pennsylvania State University
BETH E. ALLEN, University of Minnesota
MARION BAUMGARDNER, Purdue University
ANNE W. BRANSCOMB, Harvard University
JOLIE A. CIZEWSKI, Rutgers University
MARTIN W. DUBETZ, Washington University
GERALD R. FAULHABER, University of Pennsylvania
JOANNE I. GABRYNOWICZ, University of North Dakota
PAUL H. GINSPARG, Los Alamos National Laboratory
WILLIAM E. GORDON, Rice University
RICHARD E. HALLGREN, American Meteorological Society
DONALD W. KING, King Research
MICAH I. KRICHEVSKY, Bionomics International
THOMAS F. MALONE, Sigma Xi
JERRY M. MELILLO, Marine Biological Laboratory (through May 1996)
JEROME H. REICHMAN, Vanderbilt University
B.K. RICHARD, TRW
ETHAN J. SCHREIER, Space Telescope Science Institute
DIETER SÖLL, Yale University
JACK H. WESTBROOK, Brookline Technologies
RONALD L. WIGINGTON, American Chemical Society (retired)

National Research Council Staff

PAUL F. UHLIR, Study Director
JULIE M. ESANU, Program Associate
DAVID BASKIN, Project Assistant
ALEXANDRA C. SPAITH, Project Assistant

U.S. NATIONAL COMMITTEE FOR CODATA

DAVID R. LIDE, JR., National Institute of Standards and Technology (retired), *Chair*
STANLEY M. BESEN,* Charles River Associates, Inc.
LOIS BLAINE, American Type Culture Collection
WILLIAM BONNER, University Corporation for Atmospheric Research
EVAN BUCK, Union Carbide Corporation
NAHUM D. GERSHON, MITRE Corporation
ALI GHOVANLOU, Department of Energy
BRUCE GRITTON, Monterey Bay Aquarium Research Institute
JULIAN HUMPHRIES, Cornell University
MICAH I. KRICHEVSKY, Bionomics International
DAVID MARK, State University of New York at Buffalo
GOETZ OERTEL, Association of Universities for Research in Astronomy
STANLEY RUTTENBERG,* National Center for Atmospheric Research (retired)
PAMELA SAMUELSON,* University of California at Berkeley

National Research Council Staff

PAUL F. UHLIR, Director
JULIE M. ESANU, Program Associate
DAVID BASKIN, Project Assistant
ALEXANDRA C. SPAITH, Project Assistant

---

* Term ended June 30, 1996.

COMMISSION ON PHYSICAL SCIENCES, MATHEMATICS, AND APPLICATIONS

ROBERT J. HERMANN, United Technologies Corporation, *Co-Chair*
W. CARL LINEBERGER, University of Colorado, *Co-Chair*
PETER M. BANKS, Environmental Research Institute of Michigan
LAWRENCE D. BROWN, University of Pennsylvania
RONALD G. DOUGLAS, Texas A&M University
JOHN E. ESTES, University of California at Santa Barbara
L. LOUIS HEGEDUS, Elf Atochem North America, Inc.
JOHN E. HOPCROFT, Cornell University
RHONDA J. HUGHES, Bryn Mawr College
SHIRLEY A. JACKSON, U.S. Nuclear Regulatory Commission
KENNETH H. KELLER, University of Minnesota
KENNETH I. KELLERMANN, National Radio Astronomy Observatory
MARGARET G. KIVELSON, University of California at Los Angeles
DANIEL KLEPPNER, Massachusetts Institute of Technology
JOHN KREICK, Sanders, a Lockheed Martin Company
MARSHA I. LESTER, University of Pennsylvania
THOMAS A. PRINCE, California Institute of Technology
NICHOLAS P. SAMIOS, Brookhaven National Laboratory
L.E. SCRIVEN, University of Minnesota
SHMUEL WINOGRAD, IBM T.J. Watson Research Center
CHARLES A. ZRAKET, MITRE Corporation (retired)

NORMAN METZGER, Executive Director
PAUL F. UHLIR, Associate Executive Director

The National Academy of Sciences is a private, nonprofit, self-perpetuating society of distinguished scholars engaged in scientific and engineering research, dedicated to the furtherance of science and technology and to their use for the general welfare. Upon the authority of the charter granted to it by the Congress in 1863, the Academy has a mandate that requires it to advise the federal government on scientific and technical matters. Dr. Bruce Alberts is president of the National Academy of Sciences.

The National Academy of Engineering was established in 1964, under the charter of the National Academy of Sciences, as a parallel organization of outstanding engineers. It is autonomous in its administration and in the selection of its members, sharing with the National Academy of Sciences the responsibility for advising the federal government. The National Academy of Engineering also sponsors engineering programs aimed at meeting national needs, encourages education and research, and recognizes the superior achievements of engineers. Dr. William A. Wulf is president of the National Academy of Engineering.

The Institute of Medicine was established in 1970 by the National Academy of Sciences to secure the services of eminent members of appropriate professions in the examination of policy matters pertaining to the health of the public. The Institute acts under the responsibility given to the National Academy of Sciences by its congressional charter to be an adviser to the federal government and, upon its own initiative, to identify issues of medical care, research, and education. Dr. Kenneth I. Shine is president of the Institute of Medicine.

The National Research Council was established by the National Academy of Sciences in 1916 to associate the broad community of science and technology with the Academy's purposes of furthering knowledge and advising the federal government. Functioning in accordance with general policies determined by the Academy, the Council has become the principal operating agency of both the National Academy of Sciences and the National Academy of Engineering in providing services to the government, the public, and the scientific and engineering communities. The Council is administered jointly by both Academies and the Institute of Medicine. Dr. Bruce Alberts and Dr. William A. Wulf are chairman and vice chairman, respectively, of the National Research Council.

Preface

Science is perhaps the most truly international of human enterprises, not because it is based on cooperation among nations but rather because its practitioners collaborate and compete in an endeavor in which nationality tends to be irrelevant. Since Galileo corresponded with Kepler and Leeuwenhoek sent his observations to the Royal Society in London, the internationalization of science has grown steadily, with post-World War II travel and fellowship support, growing numbers of opportunities for academic exchanges, and, most recently, electronic information exchange and the end of the Cold War. Financial support for science, especially by governments, has been crucial in enabling this international interchange.

At the same time, such support has made possible a manyfold expansion of science, in numbers of scientists, institutions, and publications. The amount of scientific data has grown both because there are more scientists and because with technological advances, each of them can produce more data than ever before.

But as new technologies and working styles rapidly supersede older ones, altogether new problems are appearing that affect how data are handled and used. Some of these are technical; others are directly related to the substance of the science itself. Still others involve legal, economic, and social dilemmas that arise when the work of scientists is integrated into the daily life of the larger society. One such issue that is particularly important for science concerns the exchange of scientific data, especially exchange across national boundaries. Factors affecting exchange of data in the natural sciences, and the significance of important changes affecting access to data, are the central topic of this report.

In 1994 the U.S. National Committee for the Committee on Data for Science and Technology (CODATA),[1] organized under the Commission on Physical Sciences, Mathematics, and Applications of the National Research Council, established the Committee on Issues in the Transborder Flow of Scientific Data to investigate the changing environment for the international exchange of scientific data in the natural sciences. The results of the study committee's deliberations constitute the substance of this report. Its aim is to examine the current state of global access to scientific data, to identify strengths, problems, and challenges that exist today or appear likely to arise in the next few years, and to recommend actions to build on those strengths and ameliorate or avoid those problems. The focus is on data in the natural sciences, because that is the primary subject-matter purview of CODATA, but this should not be interpreted as implying that the committee considers engineering or social science data to be less important. Although many of the issues identified in this study pertain to those other discipline domains as well, they involve different contexts and problems that require additional study.

The committee included practicing scientists who both contribute to and use the data resources needed in international scientific efforts, computer scientists and engineers who create and maintain the means for such exchange, economists who interpret how the scientific enterprise sustains itself in the larger society, and lawyers who specialize in scientific data problems. Learning to talk with each other and to understand different perspectives was a necessary step toward reaching some agreement on important issues and crafting recommendations to address them. In all of these efforts, the focus was on understanding how to ensure global access to the data required to conduct basic research in the natural sciences. The committee's specific recommendations are presented in the relevant sections of the main text and are listed together in the summary that begins this report.

R. Stephen Berry, *Chair* Paul F. Uhlir, *Study Director*
Committee on Issues in the
 Transborder Flow of Scientific Data

---

[1]CODATA is an interdisciplinary committee of the International Council of Scientific Unions. CODATA is concerned with all types of quantitative and qualitative data resulting from experimental measurements or observations in the physical, biological, geological, and astronomical sciences. Particular emphasis is given to data management problems common to different scientific disciplines and to data used outside the field in which they were generated. The general objectives are the improvement of the quality and accessibility of data, as well as the methods by which data are acquired, managed, and analyzed; the facilitation of international cooperation among those collecting, organizing, and using data; and the promotion of an increased awareness in the scientific and technical community of the importance of these activities. Additional information about CODATA is available on-line at <http://www.cisti.nrc.ca/programs/codata/welcome.html> or from the CODATA Secretariat, 51 Boulevard de Montmorency, 75016 Paris, France.

Acknowledgments

The study committee is very grateful to the many individuals who played a significant role in the completion of this study. The committee held its first meeting on January 20-22, 1995, and extends its thanks to the following individuals who provided briefings and other information from the study's sponsoring agencies: Gerald Barton and Christopher Miller of the National Oceanic and Atmospheric Administration; William Blanpied of the National Science Foundation; Joseph Bredekamp of the National Aeronautics and Space Administration; Kurt Molholm of the Defense Technical Information Center; John Rumble of the National Institute of Standards and Technology; and Elliot Siegel of the National Library of Medicine. At this meeting, the committee also received background briefings from other federal agency representatives, including Gesina C. (Cynthia) Carter, Wanda Farrell, and Jay Snoddy of the Department of Energy; Paul Kanciruk of the Oak Ridge National Laboratory; and Dorothy Bergamaschi of the Department of State.

The committee also extends its thanks to the issue area experts who briefed the committee at its September 22-24, 1995, meeting: Gershon Sher of the National Science Foundation, Gregory van der Vink of the Incorporated Research Institutions for Seismology, Carlos Gamboa of the Pan American Health Organization, Amy Gimble of the American Association for the Advancement of Science, and Lane Smith of the U.S. Agency for International Development participated in a roundtable discussion of data access issues in less developed countries; Hal Varian of the University of California at Berkeley discussed economic aspects of data transfer on the Internet; John Baumgarten of Proskauer, Rose, Goetz, and Mendelsohn, Mary Levering of the Library of Congress, Edward Malloy of the State Department, Pamela Samuelson of the University of California at Berkeley, and Peter Weiss of the Office of Management and Budget participated in a discussion concerning intellectual property issues; and Richard Greenfield of the National Science Foundation provided an update of the World Meteorological Organization's data exchange policy.

The committee is very appreciative, as well, of the contributions of the more than 200 individuals who responded to the committee's "Inquiry to Interested Parties" (see Appendix D of this report)[1] and of the many data experts who responded to specific requests for information used in the body of the report. In addition, the committee would like to express its gratitude to the members of the U.S. National Committee for CODATA, who provided oversight of this study, and the following ex officio members, who provided liaison with other relevant activities: David R. Lide, Jr., consultant and chair of the U.S. National Committee for CODATA; Francis Bretherton, of the University of Wisconsin, chair of the National Research Council's Committee on Geophysical and Environmental Data; and Ferris Webster, of the University of Delaware, chair of the CODATA Working Group on Data Access. The committee also wishes to acknowledge the assistance of Anne Linn, director of the Committee on Geophysical and Environmental Data, and Wendy White, director of the Committee on International Organizations and Programs.

Finally, the committee would like to recognize the invaluable contributions of the National Research Council staff without whom this report could not have been completed: Paul F. Uhlir, associate executive director of the Commission on Physical Sciences, Mathematics, and Applications, who served as study director; Julie M. Esanu, for the program and research assistance provided to the committee; David Baskin and Alexandra Spaith for the staff support to the committee; and editorial consultant Roseanne Price, who edited the final manuscript.

---

[1]An edited compilation of the responses to the committee's "Inquiry to Interested Parties" is available on the U.S. National Committee for CODATA's World Wide Web site at <http://www.nas.edu/cpsma/codata.htm>.

Contents

SUMMARY   1

1 INTRODUCTION   17
  Charge and Scope of Study, 19
  Underlying Assumptions and Concerns, 21
  Notes, 22

2 TRENDS AND ISSUES IN INFORMATION TECHNOLOGY   24
  Overview of Technical Trends, 24
  Specific Technical Concerns, 35
  Data Access Issues in Developing Countries, 40
  Recommendations on Issues in Information Technology, 43
  Notes, 44

3 SCIENTIFIC ISSUES IN THE INTERNATIONAL EXCHANGE
  OF DATA IN THE NATURAL SCIENCES   47
  Types of Data and Their Use in Different Disciplines, 49
  Data Trends, Opportunities, and Challenges in the Natural Sciences, 57
  Discipline-Specific Data Issues, 68
  Access to Scientific Data in Developing Countries, 90
  Recommendations on Data Issues in the Natural Sciences, 100
  Notes, 103

4   DATA FROM PUBLICLY FUNDED RESEARCH—
    THE ECONOMIC PERSPECTIVE   110
    The Trend Toward Markets: Good or Bad for Science?, 111
    Determinants of the Structure of Scientific Data Distribution, 114
    Privatization: When Does It Make Sense?, 116
    Pricing Publicly Funded Scientific Data, 124
    Electronic Access and Internet Congestion, 126
    Recommendations Regarding Economic Aspects of Scientific Data, 128
    Notes, 129

5   THE TREND TOWARD STRENGTHENED INTELLECTUAL
    PROPERTY RIGHTS: A POTENTIAL THREAT TO
    PUBLIC-GOOD USES OF SCIENTIFIC DATA   132
    Salient Features of the Predigital Status Quo, 135
    Digital Technology—Disrupting the Balance of Public and
       Private Interests, 139
    The Drive for Legal Protection of Noncopyrightable Databases, 145
    Charting a Well-Considered Course in the New Era, 161
    Recommendations Regarding Legal Developments Affecting
       Access to Data, 171
    Notes, 172

APPENDIXES

A   ABBREVIATIONS AND ACRONYMS   191
B   GLOSSARY   196
C   EXAMPLES OF SUCCESSFUL INTERNATIONAL DATA
    EXCHANGE ACTIVITIES IN THE NATURAL SCIENCES   205
D   INQUIRY TO INTERESTED PARTIES ON ISSUES
    IN THE TRANSBORDER FLOW OF SCIENTIFIC DATA   220

INDEX   227

*This study is dedicated in fond memory of
Gesina C. (Cynthia) Carter,
director of the U.S. National Committee
for CODATA
from 1978 to 1991.*

Summary

In today's technological world, sustaining science as a source of new knowledge and innovation has become as important to modern society as maintaining the nation's capabilities in manufacturing, trade, and defense. The extent to which public funding in the developed world supports science is testimony to society's recognition that basic as well as applied research must be carried out to advance the public interest.

Science itself is a living enterprise. With few exceptions, acquisition of scientific knowledge is a cumulative process that depends on researchers' continuing ability to collect and share data. This capability has been strengthened by the advent of information technology, which is supplying powerful new tools and enabling new styles of working. However, far-reaching changes involving complex technical, economic, and legal issues also have begun to alter the conditions for exchange of data among scientists, especially across national boundaries.

To help understand the impact of such changes and to learn what actions are needed to ensure full and open exchange of scientific data[1] worldwide among researchers in the natural sciences, the Committee on Issues in the Transborder Flow of Scientific Data undertook a study responding to the following charge:

• Outline the needs for access to data in the major research areas of current scientific interest that fall within the scope of CODATA—the physical, astronomical, geological, and biological sciences.
• Characterize the legal, economic, policy, and technical factors and trends that have an influence—whether favorable or negative—on access to data by the scientific community.

- Identify and analyze the barriers to international access to scientific data that may be expected to have the most adverse impact in discipline areas within CODATA's purview, with emphasis on factors common to all the disciplines.
- Recommend to the sponsors of the study approaches that could help overcome barriers to access in the international context.

This study addresses issues in effective access to data in numerical, symbolic, and image forms by scientists for scientific research purposes, rather than to bibliographic or purely textual information. The focus is on digital rather than analog data, since practically all scientific data are now collected and stored digitally, and most older data are being transferred to digitized electronic formats. The scope of inquiry also is limited to data in the natural sciences, which is the principal subject-matter focus of CODATA.[2]

Because the sponsors of the study are U.S. federal government science agencies, the committee has emphasized those trends, issues, and barriers that have an impact on international access to data collected and used in publicly funded, basic research programs—that is, scientific research conducted as a public good. Despite this emphasis, the committee took into account the continua between fundamental and applied research, between raw data and processed information, and between public and private uses of scientific data. Indeed, the most vexing public policy issues facing the international scientific community in the exchange of data involve defining the appropriate balance of divergent interests.

Underlying the committee's approach, however, and informing its conclusions and recommendations, is the principle that full and open exchange of scientific data—the "bits of power" on which the health of the scientific enterprise depends—is vital for advancing the nation's progress and for maximizing the social benefits that accrue from science worldwide.

COMPLEX DEVELOPMENTS AFFECTING ACCESS TO SCIENTIFIC DATA

Recent Trends and Emerging Concerns

Freedom of inquiry, the full and open availability of scientific data on an international basis, and the open publication of results are cornerstones of basic research that U.S. law and tradition have long upheld. For many decades, the United States has been a leader in the collection and dissemination of scientific data, and in the discovery and creation of new knowledge. By sharing and exchanging data with the international community and by openly publishing the results of research, all countries, including the United States, have benefited. Today, however, many rapid changes portend significant consequences, some possibly adverse, for the conduct of basic research in the natural sciences.

In broad terms, the challenges of greatest import for full and open global sharing of scientific data are those associated with two quite recent trends:

1. The need for scientists to adapt to conducting research with data that come in rapidly increasing quantities, varieties, and modes of dissemination, frequently for purposes far more interdisciplinary than in the past; and
2. The worldwide trend toward imposition of increasing economic and legal restrictions on access to scientific data gained from publicly funded research.

The former obliges scientists to reexamine how they carry out their calling. The latter impels the scientific community to become more involved in understanding the significance of public policies and legislative activities that can have a profound impact on their work.

Chief among recent developments affecting access to scientific data is the widespread use of powerful new technologies for data acquisition, storage, and communication, as well as their inevitable consequence, the rapidly growing quantity of data that scientists are generating, preserving, and distributing. Moreover, because of increasingly diverse applications for the results of scientific research, these data are becoming ever more useful and valuable in many sectors outside the specific areas of research that generate them. Finding ways to distribute such information to all who want it—equitably, reliably, and in keeping with the principle of full and open exchange as a sine qua non of progress in science—is the greatest challenge this committee identified while conducting its study.

Although scientific interchange was an important stimulus for development of the Internet and initially represented one of its greatest uses, commercial activities and entertainment now far surpass scientific use of the network and may be expected to dominate policymaking for the electronic exchange of information. This development raises questions about the scientific community's continuing capability to utilize what has clearly become a beneficial and versatile tool for scientific exchange and interaction. The economic framework for a global information system and legal models for dealing with conflicting interests are increasingly influenced by stakeholders who have no long-term responsibilities for, or concern about, sustaining publicly funded scientific inquiry. Simultaneously, the government science agencies expected to assume long-term responsibilities for sustaining scientific inquiry are questioning their capacity to continue to invest at traditional levels in the creation, preservation, and dissemination of scientific data.

Issues in Information Technology

Some technical trends and developments have had a significant, largely positive impact on the management and international exchange of scientific data. These include the steadily decreasing cost of computing and communication;

greatly enhanced capabilities for collecting scientific data, for example, from remote sensors; increasing exploitation of broadband networks and capabilities for transmission of video data over networks; the advent of digital wireless communication; increasing support for collaborative work by long-distance communication; growing capabilities for natural language processing; increasing recognition of the importance of standards in data structures and in networked communication; growing acceptance of the need for cooperation in monitoring and controlling network activity; and increasing use of intranets.

Associated with advances in, and increasing reliance on, information tools and infrastructure are a number of problems that present barriers to access, including the growing congestion of the Internet and consequent constraints on scientific communication and research; the storage and distribution of data that are inadequately described or indexed for significant numbers of potential users; the rapid obsolescence of electronic information-processing tools and storage media; the vulnerability of electronic networks and data repositories to accidental or deliberate damage; and the growing competition for use of currently limited network resources. Another difficulty—the current lack of adequate access to scientific data in developing countries—nevertheless has the potential to improve quickly.

Data Issues in the Natural Sciences

The natural sciences—including the physical, astronomical, geological, and biological sciences—face a number of trends, opportunities, and challenges affecting researchers' capabilities for sharing data. The most obvious involves dealing with the exponentially growing volume of accumulating scientific data, which now, as a result of expanding computational power, also includes elaborate simulations that often incorporate animation as well as quantitative information. With the end of the Cold War has come declassification of some data that are now providing many new opportunities for researchers, particularly in the Earth sciences. In addition, because of the breadth and scale of major interdisciplinary, global-scale research efforts such as the International Geosphere-Biosphere Programme, the Human Genome Project, and the Hubble Space Telescope project, data from individual disciplines have become important to understanding and progress in other fields. Making data available, comprehensible, and useful across disciplinary boundaries has become a far greater imperative than before these projects existed. This task, however, is complicated by the fact that scientific data do not constitute a uniform, easily accessible body of information.

For example, scientific data may be categorized in many ways: by form or coding (numeric, symbolic, still image, animation, or other); by content; by means of generation; by level of quality and complexity; by the source of support for the data-accumulating activity; by time and space, in the case of observational, geospatial records; and by the institutional structures through which the data are distributed and stored. Certain of these characteristics, such as level of quality (including degree of review and certification) and institutional origin, have given rise to additional complications associated with the increasingly pervasive electronic distribution of scientific data.

Some data issues are more discipline specific. Perennial problems affecting access to data in the observational sciences, for example, include gaps in quality control, incompatibility of data streams, inadequate documentation of data sets, and difficulty in meeting the requirements for long-term retention of data. In the biological sciences, the variety of attributes and qualifiers included with each observation and differences in terminology and usage put a heavy burden on any supplier of data to identify and specify the character of the data precisely enough to prevent misinterpretation. In the laboratory physical sciences, as in many other branches, fragmentation of data into numerous, autonomous, and often incompatible databases with different formats and levels of quality is a chronic problem.

Putting scientific data to use rapidly in sectors outside the immediate discipline of origin poses additional challenges to the longer-term effort to provide full and open access. In the observational environmental sciences, for instance, massive archives and reliable institutional memory are necessary to keep the data accessible and intelligible. Simultaneously, however, data also must be available to meet the public's need for warnings of natural hazards and disasters and for commercial use by the private sector. In addition, availability of data can be affected by governmental concerns related to national security, foreign policy, and international trade. Newly adopted or proposed restrictions on previously open and unrestricted data have caused particular concern in the Earth science communities, for example.

Another significant concern regarding full and open access to scientific data is related to commercialization of electronic publication and electronic databases. Science operates according to a "market" of its own, one that has rules and values different from those of commercial markets. While protection of intellectual property may concern a scientist who is writing a textbook, that same scientist, publishing a paper in a scientific journal, is motivated by the desire to propagate ideas, with the expectation of full and open access to the results. To commercial publishers (including many professional societies), protection of intellectual property means protection of the rights to reproduce and distribute printed material. To scientists, protection of intellectual property usually signifies assurance of proper attribution and credit for ideas and achievements. Generally, scientists are more concerned that their work be read and used rather than that it be protected against unauthorized copying. These conflicting viewpoints pose challenging problems for science and the rest of society. Current discussions are seeking a balance between protecting publicly supported activities that advance the public welfare and strengthening individual rights to intellectual property.

Associated with the internationalization of scientific data collection and use has been the growth of data centers—dedicated, stable institutions supporting collaborative data sharing across international boundaries and providing verification, documentation, archiving, and dissemination of large, accumulating data sets. The scientific community is increasingly dependent on these data centers—on their skills in data management and distribution and on their capacity to support international scientific efforts.

Finally, an important concern in global access to scientific data is the need to improve capabilities for electronic communication by researchers working in developing countries. A two-way communication capability is needed: scientists in developing countries, like scientists everywhere, generate data that are just as important to science as the data they acquire. Finding ways to help less developed nations acquire affordable electronic network services is an effort that can and should be undertaken by concerned national and international organizations with the help of the telecommunications sector.

The constraints caused by inequalities among nations in access to scientific data are especially damaging to those sciences concerned with inherently international issues, such as food production, biodiversity, the prevention and cure of communicable diseases, global climate change, and other Earth system processes. Each of these sciences requires the generation of globally compatible, accessible, and usable data sets related to terrestrial ecosystems, the physical environment, and human activities. Collaboration among members of the scientific communities in every nation, rich and poor, in developing global observational data sets and in ensuring the subsequent full and open availability of those data is imperative; its importance cannot be emphasized too strongly.

Economic Aspects of Scientific Data

As the quantities and uses of scientific data have expanded, and as nations' discretionary budgets have become increasingly constrained, some governments have begun to privatize activities previously delivered by the public sector and have sold some products and services on a commercial basis—including the generation and distribution of scientific data. This development has stimulated fears that scientific data may become priced beyond the means of the scientific communities, even in the more developed countries, despite the fact that the conduct of basic scientific research, like other government activities related to public health and safety, serves the public welfare and thus is appropriately supported by government funding.

Although economists may initially see privatization as a positive development for science, careful analysis suggests that a market model different from that of ordinary commerce is more appropriate for scientific activity for several reasons. First, the conduct of some scientific research is itself tightly tied to the collection, maintenance, and distribution of the data generated by that research. In particular, in the observational sciences, whose databases can be massive, separating the gathering, archiving, and maintenance of data from their distribution is likely to be more costly and inefficient than keeping them integrated.

Second, the contributors of scientific data, particularly in basic research, are frequently also the consumers of such data, and nonmonetized exchange of data may be most efficient in such cases. Third, in many situations, the market for scientific data is not large enough to support more than a single commercial supplier, if that. Finally, most basic research is necessarily funded from public sources. Privatizing the distribution of those data would mean that the funds now provided in grants to institutions supplying data would be channeled instead (if such funds were still available) to grants to individual scientists as users of data. Such funds in small grants to individuals are likely to be vulnerable to even the slightest budgetary pressure, thus potentially compromising the long-term health of science. Direct appropriation or block grant support to institutions with broad responsibilities for data management, preservation, and distribution, while not assured of continuity, is typically more stable and secure and is fortified by institutional memory that recognizes and supports the continued utility of archived data.

At issue now is whether or when the government should remove itself entirely as a distributor of scientific data. (There is no question here regarding the continued support by government of data generation; it is a part of the process of doing basic research that falls outside the charge of this study.) Largely because of the possibility of monopoly control and the potential threat to the principle of full and open availability of data, the government should not remove itself as a primary distributor of the scientific data that its funding has produced, without adequate safeguards as discussed below.[3]

The concern that privatization, accompanied by high prices and legal restrictions, would limit scientists' access to data needed for their work is paralleled by a similarly serious concern among economists about the possibilities for unrestricted monopolization, particularly by any party whose objectives do not include advancing the public interest. Whether they are private or governmental in nature, profit-making monopolies would endanger science, whereas privatization structured so as to encourage competition in supplying value-added data to multiple user communities could well represent good public policy.

Any pricing policies that bear on the availability of scientific data should reflect this information's characteristics as a public good—a resource that is both nondepletable (cannot be diminished by repeated use) and indivisible and nonexcludable (once having been supplied to some, cannot easily be denied to others). Because there is no social cost from repeated use, price differentiation may be justified in many situations, to ensure that the needs of the scientific community are met. Pricing of government-funded data in a differentiated system should ensure that data are available at no cost to those who provide them or otherwise contribute substantively to any given data set; for others, including commercial users, prices for data should cover the costs of serving those users. Because there is a cost associated with repeated distribution, marginal pricing has been the policy in many of the sciences. It allocates the smallest nonzero cost to users and thus is consistent with the principle of full and open exchange of data.

Internet congestion, a growing problem for transnational exchange of scientific data, has obvious economic aspects and will be resolved only if participating nations and network providers work together. For the scientific community, a partial solution may involve the creation of separate intranets.

**Intellectual Property Rights in Data:
Legal Constraints on Full and Open Access?**

The emergence of a new intellectual property rights model that protects the contents of electronic databases as well as those in print has the potential to significantly affect the international flow of scientific data. The problem has reached a crux with the current attempts, national and international, to establish a legal framework that threatens to subordinate the needs of data users working in the public interest to the desires of those seeking protection of investments in creating and maintaining databases. Unfortunately, and until very recently, the input into this legislative process at all levels by the scientific and educational communities has been all but nonexistent. Sustained action by those sectors is needed to avert possible restrictions on the full and open exchange of scientific data.

The U.S. Constitution articulates the legal protection of technological inventions and of literary and artistic works through the patent and copyright systems, which attempt to balance incentives to create against the public interest in free competition. Any publicly disclosed technology or information that does not meet the eligibility requirements for protection under U.S. patent and copyright laws becomes public domain matter that anyone can appropriate freely. Moreover, the special needs of libraries, educators, and researchers for access to the copyrighted literature has been recognized under the concept of fair use.[4]

But this traditional balancing of private and public rights has become more complex in the information age. Many information goods with commercial value, notably the contents of most electronic databases, are not eligible for patent or copyright protection, and database producers consequently face the threat of rapid duplication by free-riding competitors who do not contribute to the costs of collecting, managing, or disseminating the relevant data. In its 1991 decision in *Feist Publications, Inc. v. Rural Telephone Service Co.*,[5] the U.S. Supreme Court raised the threshold of eligibility for copyright protection, requiring significant original and creative authorship in the selection and arrangement of contents and not simply industrious compiling efforts. Earlier, the Commission of the European Communities (CEC) had started to develop a new protection framework for databases to encourage their commercialization in Europe. This culminated in the formal adoption of a new European Directive on Databases by the CEC in March 1996, which reflected influences by the *Feist* decision, as well as other concerns in Europe. In May 1996, legislation similar to the final European Directive, but even more protective, was introduced in the U.S. House of Representatives (H.R. 3531), and in August 1996, a proposal almost identical to

SUMMARY   9

the proposed U.S. legislation was placed before a Diplomatic Conference under the auspices of the World Intellectual Property Organization (WIPO) with a view to adopting a new protocol to the Berne Convention that would protect non-copyrightable databases in a tailor-made legal regime. Action on this proposal has been postponed until later in 1997.

Scientific data already largely compiled and distributed in electronic form constitute one of many types of data and information that will be affected by the legal framework now evolving in response to conflicting needs. Although new forms of legal protection may be needed to attract private investment to finance the creation and maintenance of electronic databases, including those for use in science and technology, current European and U.S. initiatives would confer a monopoly on database developers far broader and stronger than is needed to avert market failure. The pending legislation would create exclusive, monopolistic property rights of virtually unlimited duration, but without public policy limitations. If adopted in their current form, these legal proposals could jeopardize basic scientific research and education, eliminate competition in the markets for value-added products and services, and raise existing thresholds to entry into insuperable legal barriers to entry.

If put into practice, such measures could restrict the full and open access to data on which scientists and educators have depended. Neither the already adopted European Directive on Databases nor the proposed WIPO protocol and pending U.S. legislation would provide adequate fair use safeguards that recognize the needs of the scientific and educational communities for unrestricted access to data at affordable prices. They take little or no cognizance of the public-good character of scientific data for research and educational purposes.

More generally, such an approach ignores the contribution of basic science to the ability of U.S. firms to predominate in markets for technology and information goods. Despite a general consensus on the need for sustained levels of investment in research and development, the proposed database laws could change the status quo—without anyone's wanting it to happen—by elevating the price of the one raw material to which U.S. researchers have always had ready access. If less available scientific information were to translate to fewer applications of economic importance, the end result would be a loss of U.S. technological competitiveness in an integrated world market.

It is therefore essential to retain a "fair use" zone in cyberspace and in other media to protect the strong public interest in ensuring that certain uses and certain users, including the scientific and educational communities, are neither priced out of the market nor forced to cut back the basic research that has played a crucial role as a public good in the economic and technological growth of the United States. The pending legislative proposals, which the committee considers to be precipitous and radical attempts to alter the terms and conditions under which scientific data may be accessed and used on a worldwide basis, have the potential to do severe damage to the scientific enterprise. The scientific community and its defenders must step in quickly to insist on further, open debate before these changes reach implementation.

RECOMMENDATIONS

General Guideline

Based on its deliberations and understanding of the issues involved, the committee believes that the following overarching principle should guide all policy decisions concerning the management and international exchange of scientific data in the natural sciences: *The value of data lies in their use. Full and open access to scientific data should be adopted as the international norm for the exchange of scientific data derived from publicly funded research. The public-good interests in the full and open access to and use of scientific data need to be balanced against legitimate concerns for the protection of national security, individual privacy, and intellectual property.*

Recommendations on Data Issues in the Natural Sciences

1. Governmental science agencies and intergovernmental organizations should adopt as a fundamental operating principle the full and open exchange of scientific data. By "full and open exchange" the committee means that the data and information derived from publicly funded research are made available with as few restrictions as possible, on a nondiscriminatory basis, for no more than the cost of reproduction and distribution.

2. The International Council of Scientific Unions (ICSU), together with the scientific Specialized Agencies of the United Nations, the Organisation for Economic Co-operation and Development Megascience Forum, and the national science agencies and professional societies of member countries, should consider developing a distributed international network of data centers. Such a network should draw on the strengths of successful examples of international data exchange activities as described in Appendix C of this report, including, in particular, the ICSU World Data Centers, and become a prominent part of the global information infrastructure that has been proposed by the "Group of Seven" nations. To facilitate the international dissemination and interdisciplinary use of scientific data, all public scientific data activities, including the network of data centers, should plan for and commit to providing human and financial resources sufficient for carrying out the following functions:

    a. Involve experts from the relevant disciplines, together with information resource managers and technical specialists, in the active management and preservation of the data;

b. Develop and maintain up-to-date, comprehensive, on-line directories of data sources and protocols for access;

c. Provide documentation (metadata) adequate to ensure that each data set can be properly used and understood, with special attention given to making the data usable by individuals outside the core discipline area. This problem is particularly acute within the biological sciences, in which imprecision and variations in taxonomic definitions and nomenclature pose significant barriers to communication, even among the biological subdisciplines. The committee suggests that the CODATA Commission on Standardized Terminology for Access to Biological Data Banks be enhanced into a true international consultative body and that similar mechanisms be developed for other disciplines, as needed;

d. Incorporate advances in technology to facilitate access to and use of scientific data, while overcoming incompatibilities in formats, media, and other technical attributes through vigorous coordination and standardization efforts;

e. Institute effective programs of quality control and peer review of data sets; and

f. Digitize all key historical data sets and ensure that every important condition for the long-term retention of data be met, including the adoption of appropriate retention and purging criteria and the timely transfer of all data sets to new media to prevent their deterioration or obsolescence.

3. The ICSU and other professional scientific societies should encourage the study of, and publication of peer-reviewed papers on, effective data management and preservation practices, as well as promote the teaching of those practices in all institutions of higher learning.

4. All scientists conducting publicly funded research should make their data available immediately, or following a reasonable period of time for proprietary use. The maximum length of any proprietary period should be expressly established by the particular scientific communities, and compliance should be monitored subsequently by the funding agency.

5. As a corollary to recommendation 2.a above, publicly funded scientific databases should be maintained either directly or under subcontract by the government science agencies with the requisite discipline mission and need. In the United States, the Office of Science and Technology Policy should develop an overall policy for the long-term retention of scientific data, including a contingency plan for protecting those data that may become threatened with the loss of their institutional home.[6]

6. With regard to improving access to scientific data in developing countries, the committee makes the following recommendations:

a. International development organizations, together with professional societies, should provide targeted training programs for scientists in the use of computers, with emphasis on the management of digital data in specific disciplines.

b. Foreign aid agencies should (i) make available to individual scientists in developing countries more direct, peer-reviewed grants that include support for access to data, and (ii) facilitate the involvement of scientists in such nations in their own countries' capacity-building initiatives, research policy decisions, and national database construction efforts.

c. Scientists in developing countries should be encouraged to organize to promote the policy of full and open access to scientific data in their own countries, as well as to make their data available internationally.

d. The ICSU, together with funding agencies and nongovernmental bodies, should strengthen its efforts to assist developing countries in undertaking their own scientific studies and encourage scientists engaged in such studies to take active roles in the international scientific community, where their efforts can be appreciated and used. Legal and procedural protocols must be developed to provide for fair and equitable sharing of any resulting intellectual property.

e. Until affordable and ubiquitous electronic network services are available, national and international scientific societies and foreign aid agencies should establish or improve their existing efforts to send extra stocks of scientific publications to libraries and research institutions in developing countries that need them.

7. Finally, the ICSU, together with the principal national and international scientific organizations mentioned in Recommendation 2 above, should convene a series of major international meetings to initiate meaningful action on these recommendations.

Recommendations on Issues in Information Technology

1. The principal scientific societies and the Internet Engineering Task Force (IETF) should begin a long-term planning effort to assess the carrying capacity and distribution capability of the Internet, using projections of storage and transmission capacity and of demand and taking into account the next generation of Internet protocols. Scientific societies should encourage their publication committees to maintain contact with the IETF and keep their members abreast of advances in technologies useful for scientific information management. One option that scientific societies and government science agencies should evaluate

is the creation of dedicated international science networks, such as the Internet II now being developed.

2. To improve the technical organization and management of scientific data, the scientific community, through the government science agencies, professional societies, and the actions of individual scientists, should do the following:

    a. Work with the information and computer science communities to increase their involvement in scientific information management;

    b. Support computer science research in database technology, particularly to strengthen standards for self-describing data representations, efficient storage of large data sets, and integration of standards for configuration management;

    c. Improve science education and the reward system in the area of scientific data management; and

    d. Encourage the funding of data compilation and evaluation projects, and of data rescue efforts for important data sets in transient or obsolete forms, especially by scientists in developing countries.

3. U.S. government science agencies, working with their counterparts in other nations, should improve data authentication and apply security safeguards more vigorously. They also should continue funding for research and development in information technologies that are important to the pursuit of science.

4. A consortium of intergovernmental and nongovernmental organizations, including the International Telecommunications Union, the World Bank, the Specialized Agencies of the United Nations, the International Council of Scientific Unions, and other concerned bodies, should mount a global effort to reduce telecommunications tariffs to scientists in developing countries through differential pricing or direct subsidy.

5. Foreign aid to developing countries in the form of computers, computer networks, and associated software, coupled with the training and resources necessary to operate and maintain those technologies, should be given high priority, on the basis of the potential for long-term socioeconomic returns. The communication systems must have adequate carrying capacity to meet growing demand.

Recommendations Regarding Economic Aspects of Scientific Data

The committee recommends that the economic aspects of facilities for storage and distribution of scientific data generated by publicly funded research be evaluated according to the following criteria:

- *Does the scientific research depend on a substantial public investment in one or more facilities that generate the data of interest?* If so, the data distribution facilities are most likely to benefit by being vertically integrated with the observational or experimental facilities themselves.
- *Does the (non-facilities-based) distributed scientific research involve coordination among researchers, possibly in different countries?* If so, then data distribution becomes a means of communication among contributing scientists, and for this community, the price of the data alone should be zero. If the distributor subsequently adds value to the data, then the price should be no higher than the marginal cost of adding value.[7]
- *Is the community of users roughly the same as the community of contributors?* If so, then data distribution should be priced at zero (or at marginal cost, if value is added). If there are many users who are not contributors, such as commercial customers, then some form of price discrimination to ensure zero or low prices to contributing scientific users, with possibly higher prices to others, may be appropriate.
- *Is the user community large enough to support more than one data distributor?* If so, then privatization of data distribution may be a viable policy option. If not, then privatization should occur only if the contractual arrangements are adequately protective of the needs of the scientific community. Necessary—but not necessarily sufficient—conditions for privatization to be desirable are as follows:

—The distribution of data can be separated easily from their generation.
—The scientific data set is used by others beyond the research community.
—It is easy to price discriminate/product differentiate between scientific users and other users, and it is easy for the government to contractually mandate low prices to scientific users for government-funded data.
—Privatization will not result in the unrestricted monopoly provision of the data.

The appropriate price ceiling for nonscientific users of scientific data generated through government research is incremental cost, as defined in the section titled "Pricing Publicly Funded Scientific Data" in Chapter 4. The price of scientific data to the contributing scientific community should be zero, or at most marginal cost.

Recommendations Regarding Legal Developments Affecting Access to Data

The new proposals supporting an overly protectionist property rights regime for the contents of databases and for on-line transmissions of data and other scientific information have reached an advanced stage of legislative consideration at both the national and the international levels. The committee believes that these legislative changes do not reflect adequate consideration of the potential negative impacts on scientific research and education and that they have been proposed for implementation at an unnecessarily precipitous pace. The committee therefore recommends that the Office of Science and Technology Policy, leaders from the science agencies and professional societies, and all those concerned with sustaining the health of the scientific enterprise should immediately take the following actions:

1. Present to all relevant legislative forums the principle of full and open exchange of scientific data resulting from publicly funded research, and clarify the importance of sustaining such exchange to the nation's future whenever these forums consider laws that would apply to exchange of scientific data.

2. Demand that national and international legislative processes now in progress slow to a rational pace, and that the deliberations become more public to allow the scientific and educational communities to present their views and concerns to lawmakers.

3. Advocate the incorporation of equivalents of "fair use" as part of any regulatory structure applying to databases as such, or to on-line storage and transmission of data and other scientific information. As a corollary, ensure that the public-good aspects of scientific data are preserved and promoted in laws and regulations governing intellectual property on the Internet and in any future electronic networked environments.

4. Work with Congress and the official U.S. representatives to the World Trade Organization and the World Intellectual Property Organization to ensure that the nation's interests in maintaining preeminence in science and technology are not undermined.

5. Pursue these issues not only within the United States, but also internationally through international scientific organizations and U.S. foreign-policy channels as they deal with trade and other agreements affecting intellectual property protection.

NOTES

1. By "full and open exchange" the committee means that data and information derived from publicly funded research are made available with as few restrictions as possible, on a nondiscriminatory basis, for no more than the cost of reproduction and distribution. This definition is adapted from a basic tenet regarding availability of scientific data in global change research. See "Policy Statements on Data Management for Global Change Research" (July 1991), Office

of Science and Technology Policy, DOE/EP-0001P, Washington, D.C., and National Research Council, Committee on Geophysical and Environmental Data (1995), *On the Full and Open Exchange of Scientific Data*, National Academy Press, Washington, D.C., p. 2.

2. Throughout this report, the term "scientific data" refers to data in the natural sciences.
3. The Landsat privatization effort, described in Chapter 4, is one example of unrestricted monopolistic data distribution under which the scientific community suffered loss of access. Nevertheless there may be situations in which the scientific community would benefit if a body of data were distributed either by a competitive set of private firms or by a single adequately constrained private source.
4. Before the electronic era, copyright evolved as a protection for authors and their assignees; under copyright, a document could be reproduced only with the approval of the copyright holder, under whatever terms that person chose. Copying machines made possible, even easy, violations of this protection. A doctrine of "fair use" then evolved to allow very limited copying by scholarly, educational, scientific, and other not-for-profit users, but not by any who would make commercial use of the copies. The fair use doctrine has become a principal protection of the right of the public—and thus of the scientific community—to have ready, low-cost access to copyrighted material; its economic and cultural justification rests on the nature of information as a public good that benefits users.
5. *Feist Publications, Inc. v. Rural Telephone Service Co.*, 111 S. Ct. 1282 (1991).
6. See the recommendations in National Research Council (1995), *Preserving Scientific Data on Our Physical Universe: A New Strategy for Archiving the Nation's Scientific Information Resources*, National Academy Press, Washington, D.C.
7. By "adding value" in this case is meant any transformation of the data beyond that necessary for scientific research that increases the value of the information for some or all potential users of the data.

1

Introduction

Basic scientific research fuels most of our nation's—and the world's—progress in science.[1] Society uses the fruits of such research to expand the world's base of knowledge and applies that knowledge in myriad ways to create new wealth and to enhance the public welfare (see Box 1.1). Yet few people understand how scientific advances have made possible the ongoing improvements that are basic to the daily lives of everyone. Fewer still are aware of what it takes to achieve advances in science, or know that the scientific enterprise is becoming increasingly international in character.

Freedom of inquiry, the full and open availability of scientific data on an international basis, and the open publication of results are cornerstones of basic research that U.S. law and tradition have long upheld. For many decades, the United States has been a leader in the collection and dissemination of scientific data, and in the discovery and creation of new knowledge. By sharing and exchanging data with the international community and by openly publishing the results of research, all countries, including the United States, have benefited. In this century's dramatic growth of scientific knowledge—an expansion motivated by a combination of forces including military, commercial, public benefit (especially health), and purely intellectual—a necessary component has been the wide availability of scientific information, ranging from minimally processed data to cutting-edge research articles in newly developing fields. This information has been assembled as a matter of public responsibility by the individuals and institutions of the scientific community, largely with the support of public funding.

Data are the building blocks of scientific knowledge and the seeds of discovery. Activities in the recording, analysis, and dissemination of data are motivated

17

> **BOX 1.1**
> **Examples of Benefits Derived from Scientific Research**
>
> New scientific understanding and its applications are yielding benefits such as the following:
>
> - Improved diagnoses, pharmaceuticals, and treatments in medicine;
> - Better and higher-yield food production in agriculture;
> - New and improved materials for fabrication of manufactured objects, building materials, packaging, and special applications such as microelectronics in the production arts;
> - Faster, cheaper, and safer transportation and communication;
> - Better means for energy production;
> - Improved ability to forecast environmental conditions and to manage natural resources; and
> - More powerful ways to explore all aspects of our universe, ranging from the finest subnuclear scale to the boundaries of the universe, and encompassing living organisms in all their variety.

today by the same forces that have impelled humans for thousands of years: curiosity to understand the natural world; desire to pass that understanding to succeeding generations; self-aggrandizement; and personal or national power.[2] Data challenge us to develop new concepts, theories, and models to make sense of the patterns we see in them. They provide the quantitative basis for testing and confirming theories and for translating new knowledge into useful applications for the benefit of society. The assembled record of scientific data is both a history of events in the natural world and a record of human accomplishment.[3] The international availability of these scientific data for fundamental research on a full and open basis and issues associated with ensuring global access are the primary concerns of this report.

Technological advances in recent years have led to an exponential increase in the amount of data collected, stored, and transmitted. New, ever more sophisticated sensors record observations on objects ranging from the smallest particles of matter to the largest objects in our known universe. It is now commonplace to control such large instruments as telescopes remotely, during the observation of an event, from a point hundreds or thousands of miles from the instrument. Satellites in orbit around Earth provide us with electrooptical observations, collecting billions of bits of data about our planet on a daily basis. Powerful machines unravel genomes to reveal the genetic code of life and help us decipher the secrets of heredity. In addition, rapid advances in computing, data processing and storage, and, most recently, in global telecommunications have given us the power to communicate and share the information produced by these remarkable observational and experimental tools, almost as quickly as it is generated. The exponential accumulation of these electronic data—these bits of power—and our expanding capacity to manipulate them are in turn changing the nature of scientific inquiry and its application to the great challenges facing mankind.

As in the past, generating data in the natural sciences is only the first step in the process of creating, organizing, and applying knowledge. Other elements of this endeavor include discovery of new principles, integration of information across disciplines, dissemination by formal and informal education, and application by many sectors of society. Today, however, larger interdisciplinary research efforts such as the International Geosphere-Biosphere Programme,[4] the Human Genome Project,[5] and other international "megascience" research programs[6] are creating new frameworks of knowledge not only about the universe and what constitutes it, but also about living organisms, human behavior, and their mutual interaction. In addition, traditional disciplinary research continues in field studies, the laboratories of individual scientists, and at large joint facilities.

Increasingly, all forms of research involve both formal and informal international scientist-to-scientist contact and exchanges of data. This increase in international collaboration is owing partly to changing political and economic conditions and also to the growing availability of electronic communication. Whether carried out on a large scale under cooperative agreements or less formally among individual researchers, these collaborations have become integral to the search for scientific understanding. Their success—as well as progress in achieving the public benefits of science—depends on the full and open availability of scientific data.

CHARGE AND SCOPE OF STUDY

The purpose of this report is to describe and develop new insights into the trends, issues, and problems that are shaping the transnational exchange of scientific data. Specifically, the Committee on Issues in the Transborder Flow of Scientific Data was charged with the following tasks:

• Outline the needs for access to data in the major research areas of current scientific interest that fall within the scope of CODATA—the physical, astronomical, geological, and biological sciences.
• Characterize the legal, economic, policy, and technical factors and trends that have an influence—whether favorable or negative—on access to data by the scientific community.
• Identify and analyze the barriers to international access to scientific data that may be expected to have the most adverse impact in discipline areas within CODATA's purview, with emphasis on factors common to all the disciplines.
• Recommend to the sponsors of the study approaches that could help overcome barriers to access in the international context.

Perhaps the most obvious aspect of this charge is its wide scope. The broad nature of the committee's inquiry precluded a comprehensive analysis of all the issues and trends in all the disciplines and across all geographic areas. Moreover, many activities beyond the sphere of science impinge on the transnational exchange of scientific data, a fact that required the committee to establish practical limits on its treatment of these topics.

This report focuses primarily on issues pertaining to scientists' effective access to data in numerical, symbolic, and image form, rather than bibliographic or purely textual data, for research in the natural sciences. However, the committee is acutely aware that distinctions among these categories of data are fading. Most of the discussion concerns digital rather than analog data, since practically all scientific data are now collected and stored digitally and most older data are being transferred to digitized electronic formats.

With regard to the needs for data in the physical, astronomical, geological, and biological sciences, the report incorporates by reference the more detailed and thorough analyses of research strategies produced in recent years by the National Research Council for the various natural sciences. The importance of data for fundamental research across these disciplines is described in a summary overview at the beginning of Chapter 3 and is highlighted in various examples throughout the report.

Because the sponsors of the study are U.S. federal government science agencies, the committee has emphasized trends, issues, and barriers that have an impact on international access to data collected and used in the context of publicly funded, basic research programs. Despite this emphasis, the committee took into account the continua between fundamental and applied research, between raw data and processed information, and between public and private uses of scientific data. Indeed, the most vexing public policy issues facing the international scientific community involve defining the appropriate balance of competing interests.

In addressing the international aspects of data exchange, the committee conducted a widely disseminated informal inquiry to develop at least an anecdotal sense of what data issues trouble the international scientific community today.[7] The issues were broadly divided into those affecting the economically most developed nations, defined as the countries belonging to Organisation for Economic Cooperation and Development, and those confronting the developing countries. The committee recognizes, however, that the developing countries encompass a wide spectrum of economic and technical capacities; illustrative examples of major issues are provided with reference to specific regions, countries, or institutions.

Finally, in its deliberations the committee discovered certain matters that were central to the subject but were not explicitly included in its charge. Although expressly requested to provide its advice to the agencies that supported this study, the committee became aware that many of the issues and barriers pertinent to global access to scientific data could only be addressed collectively by the world's international scientific community in concert with a broad range of national and international governmental and nongovernmental bodies. Therefore the committee considered it necessary to make several recommendations of broader scope in keeping with these concerns.

UNDERLYING ASSUMPTIONS AND CONCERNS

Several assumptions underlie the committee's work. The first is that international collaboration enhances scientists' capacity to better understand the natural world and thus strengthens the science base that is a source of important benefits to society. From this assumption, the rest follow.

Science is one of the most internationally cooperative of activities. Today, the improving means and ease of communication and travel, as well as their decreasing costs, have made transnational interactions a normal, daily part of carrying out scientific research. National boundaries are invisible in scientists' daily interactions, whether they are engaged in face-to-face discussion in a single laboratory or across great distances by electronic mail. Joint multinational authorship is common, and many funding institutions have encouraged such efforts. With the end of the Cold War, international collaboration can be expected to increase further.

The handling of scientific information—one of the results of such collaboration—has also changed dramatically. In fact, the distinction between "data" and "information" has itself become blurred. Data now include not only numerical data, but also symbolic data and images, and, for many scientists, textual data. Much of this convergence is the consequence of powerful electronic capabilities affecting the acquisition, storage, and exchange of scientific data. Primary data collected by a detector now frequently go directly into a computer for storage and processing before the person who generates the data ever sees them. In such an experiment, how are "primary" data to be defined? Fortunately, the integration of electronic methods has occurred so naturally that this question is unimportant to the working scientist, who might ask instead, "Is this the best way to collect and analyze the data?"

The storage and exchange of scientific data have been more problematic than their collection. Even within the particular communities that generate and initially use data, their storage and dissemination traditionally have required attention and expense. Moreover, many scientific data have value outside the community of origin. Entire institutions have evolved to provide data services, among them public and private data centers. Electronic media have changed the means, costs, capacities, and time scales associated with the handling of scientific data. Some of these changes are already well established, some are evolving, and some are still only imagined. What is certain is that change will continue and that our management and use of scientific data in 10 years will differ from current practice.

Another of the committee's assumptions is a corollary of the first and reflects what the committee believes is virtually a consensus of the global scientific community: that the most valued goal of scientists is that other scientists should learn of their work and use it. The common interests of all scientists, of science, and indeed of society in general thus are best served by as full and open an exchange of scientific information as possible, consistent with the preservation of scientists' capacity to continue their investigations. This assumption can sometimes put scientists at odds with other sectors of society, as discussion and examples in this report illustrate. Because the scientific community is not the only sector with an interest in the handling of scientific data and information, scientists need to remain involved in the current policy debate that will affect the prospects for continuing open, global access to scientific data.

This study has been motivated by a concern for ensuring the continuing strength of the scientific enterprise as a source of international well-being and progress; hence the analysis and recommendations reflect that motivation. The extent to which the committee's recommendations are adopted may require balancing this motivation against the motivations of others, whose objectives are not necessarily the same.

The chapters that follow (a) describe the information technology tools and capabilities that are transforming the handling and use of scientific data, and some of the principal impacts on data exchange arising from these technological developments; (b) summarize the underlying factors in international scientific data exchange, how scientists use data, and what data issues confront them as they carry out their research; (c) examine the economic aspects of data obtained from publicly funded research; and (d) analyze the conflicts arising from information technology's impact on the domain of intellectual property law that regulates scientists' access to data. Technical terms and acronyms are defined, and examples of successful data exchange activities given, in the appendixes.

NOTES

1. Basic, or fundamental, research may be defined as research that leads to new understanding of how nature works and how its many facets are interconnected. See John A. Armstrong, "Is Basic Research a Luxury Our Society Can No Longer Afford?," Karl Taylor Compton Lecture, Massachusetts Institute of Technology, October 13, 1993.
2. J.H. Westbrook, (1992), "A History of Data Recording, Analysis, and Dissemination," pp. 430-460 in *Data for Discovery: Proceedings of the Twelfth International CODATA Conference*, P. Glaeser, ed., Begell House, New York.
3. National Research Council (1995), *Preserving Scientific Data on Our Physical Universe: A New Strategy for Archiving the Nation's Scientific Information Resources*, National Academy Press, Washington, D.C.
4. See the International Geosphere-Biosphere Programme's World Wide Web site at <http://www.igbp.kva.se/index.html>. Note: In keeping with the subject and message of this report, the reader will find, in addition to references to texts and personal communications, many

INTRODUCTION 23

references to sites on the World Wide Web. Most of these are uniform resources locators, or URLs; a few are uniform resources names, or URNs. Although the validity of all of these Web addresses was determined at the time of publication, the reader is cautioned that URLs sometimes change, and that one of the shortcomings of the current state of electronic communication is inadequate tracking capability to lead someone from an old to a new URL when the address changes. The replacement of URLs by URNs is a likely solution to this problem in the coming years, but it has not yet happened.

5. For additional information about the Human Genome Project, see the National Human Genome Research Institute's Web site at <http://www.nhgri.nih.gov/HGP>.
6. Organisation for Economic Co-operation and Development (OECD) Megascience Forum (1993), *Megascience and Its Background*, OECD, Paris, France.
7. The inquiry is reprinted in Appendix D, and the results are summarized separately on the USNC/CODATA Web site at <http://www.nas.edu/cpsma/codata.htm>.

2

Trends and Issues in Information Technology

Advances in information technology offer unprecedented opportunities as well as new challenges in the international exchange of scientific data. Rapid improvements have led to ever greater computational speed, communication bandwidth, and storage capacity at costs within reach of even small-scale users—a trend that appears likely to continue well into the future.[1] Moreover, technical advances in satellites, sensors, robotics, and fiber-optic and wireless telecommunications are extending the range of technologies affecting the acquisition, refinement, analysis, transmission, and sharing of scientific data.

In this chapter, the committee examines some of the concerns that rapid changes and growing reliance on information technology have raised with respect to the exchange of scientific data. Table 2.1 frames some of the profound advances in technology that are having an impact on access to and exchange of scientific data and thus on research-related capabilities. The committee's overview of associated technical trends provides some context for its discussion of six barriers to and concerns regarding global access to scientific information, including access by scientists in developing countries. Its recommendations for technical improvements to facilitate the international sharing of scientific data are addressed to a range of participants.

OVERVIEW OF TECHNICAL TRENDS

The committee's discussion focuses on 10 trends (Table 2.2) that represent major forces of change in data and information technology. These trends interact with and reinforce each other, often further accelerating change and complicating

TABLE 2.1 Advances in Technologies Relevant to the Generation and Exchange of Scientific Data and Information

Technology	Relevance	Prognosis
High-density storage and memory	Capacity to deal with large volumes of data and high rates of transmission for today's science demands.	Precipitous drop in cost. Holographic and high-density optical memory technology will enter market.[a]
Encryption/authentication	Ability to protect copyright, the privacy of individuals, and data integrity.	Imbedded encryption in numerous products expected to make privacy and security applications manageable. Widespread application of public-key encryption.
Packet asynchronous transfer mode (ATM) communications	Support for high-speed, flexible transmission of video and images.	Long-term steady growth in ATM applications over high-speed fiber-optic links. Use of ATM within local area networks, competitive with other local area network (LAN) technologies (e.g., 100-Mbps Ethernet)
Sensors	Extension of the range of that which can be observed (more precision, more spectral range, higher sampling frequency, less calibration effort).	New multispectral sensors, improved resolution, smaller and more numerous satellites. Additional terrestrial applications (e.g., agriculture).
Small satellites (and inexpensive launches)	Lowering of barriers to entry for remote sensing applications.	Increased space and ground remote sensing activity. Broader array of applications.[b]
Wireless (space and ground) communications	Capability to enhance communication in remote areas or areas where post, telephone, and telegraph have limited capability/capacity.	Worldwide access to voice and high-speed data transmission within 5 years. Wireless systems filling in to meet communications needs that ongoing investments in fiber cable or wireline systems have been unable to accommodate.
High-performance computer processors	Enhanced capability for computationally intensive science activities (e.g., models, transformation of large data sets).	More expensive fabrication processes expected to cause reduction in supplier alternatives.[c] Potential for "single-electron" circuits.[d] Moore's law applying through 2005, or longer.
Robotics for exploration and for technology data transmission, from or to inaccessible places	Improved autonomy of vehicles for ocean and atmosphere studies and for planetary missions.	New frames for small submarines and pilotless aircraft (driven in part by military applications); "downhole" oil, gas, and geologic exploration; micro-electromechanical systems (MEMs) applications.[e]

(*continues*)

TABLE 2.1 Continued

Technology	Relevance	Prognosis
Hybrid analog/digital computers	Capacity for new sensing and reasoning power, helping machines to do "intelligent" work.	Potentially rapid advancement portended by breakthroughs in computation and neural science research.[f]
Language processing by computer	Assistance in getting relevant information to scientists on time. Capability for speakers of different languages to improve collaboration.	Improved filtering and organizing of information in response to the WWW information glut. Many graphical tools have capabilities to deal with special fonts and 16-bit character sets. Limited to "major" language pairs in the near term.
Database technology (including information retrieval, "knowbots")	Capability to deal with the extremely complex variety of information in natural sciences and medicine; support in organizing relevant, current information.	Object-oriented databases entering the market; massively parallel representations (e.g., Paradise at University of Wisconsin[g]). Widespread use of electronic agents to assist research.
Fiber-optic communications	Vastly increased capacity to accommodate rates of many gigabits per second.	New, erbium-based systems expected to reduce power and improve reliability.[h] Improved connectivity using undersea cables.[i]

[a]Demetri Psaltis and Fai Mok (1995), "Holographic Memories," *Scientific American*, November:70-76; Praveen Asthana and Blair Finkelstein (1995), "Superdense Optical Storage," *IEEE Spectrum*, August:25-31; Robert F. Service (1995), "Pushing the Data Storage Envelope," *Science*, July 21:299-300.

[b]K.C. Cole (1996), "NASA's Mission: Think Small," *Los Angeles Times*, January 21:1.

[c]A discussion of the increasing cost of semiconductor production capability is provided in G. Dan Hutcheson and Jerry D. Hutcheson (1996), "Technology and Economics in the Semiconductor Industry," *Scientific American*, 274 (1):54-62.

[d]Chappell Brown (1996), "Electron Switching Simplified," *EE Times*, January 8:35.

[e]MEMS are nanoscale machines. Applications include instrumentation within living tissue. See <http://mems.isi.edu/>.

[f]R. Colin Johnson (1995), "Mead Envisions New Design Era—Analog and Digital Techniques to Create a New 'Art Form'," *EE Times*.

[g]David J. DeWitt, (1994), "The Trend Toward Object Oriented DBMS," briefing to NASA. See <http://www.hq.nasa.gov/office/oss/aistr/1994_minutes.html>.

[h]George Gilder (1997), "Fiber Keeps Its Promise," *Forbes ASAP*, April 7:90-94; and Frank J. Denniston and Peter K. Runge (1995), "The Glass Necklace," *IEEE Spectrum*, October:24-27.

[i]See <http://www.teleport.com/~simoriah/scow/sub.htm> for more information on undersea cables and plans. Also, AT&T has announced support of "Africa One," a $1.9 billion project intended to link coastal countries in Africa.

application choices. Each is discussed below, and their actual or potential effect on international scientific data exchange among the member countries of the Organisation for Economic Co-operation and Development (OECD) is broadly characterized. The impact of these technical trends on access to scientific data in developing countries is also discussed.

Decreasing Cost of Computing and Communications

The cost of owning and operating increasingly powerful computers has dropped dramatically over the past several decades. Today's personal computers, for example, offer the processing speed of workstations of fewer than 5 years ago at a fraction of the cost. The availability of information technology products with ever-increasing computing, communication, and storage capability has contributed to the ubiquitous assimilation of computers into modern daily life, and complex applications taking advantage of continually improving computer performance have emerged. Among other uses, information technology is being applied increasingly to product development, manufacturing, and distribution, as well as to new financial services such as debit/credit transactions and investment portfolio management.

One effect of this phenomenon is an opportunity for "technology leapfrogging": late entrants to the use of information technology can enjoy the immediate advantage of low-cost systems, without having had to make earlier investments in more expensive and less capable technologies and then carry the burden of depreciation of that investment. Modern computing technology is thus increasingly accessible to low-budget endeavors as prices fall also to the press of mass production and competition.[2]

Even though the pace of change can be daunting to information technology newcomers, in general it should become easier and cheaper with time to obtain technology to participate in the global sharing of scientific information. In the context of the natural sciences, this means that scientists and other users in developing countries or in economically depressed regions such as those in Eastern Europe and the former Soviet Union are increasingly able to acquire new computing and communications tools for carrying out their work.

Enhanced Capabilities for Collecting Scientific Data

The natural sciences produce prodigious amounts of data. Earth observation and weather systems lead the way, with the potential for collecting terabytes[3] per day. The same trends in low-cost microelectronics that are fueling the information and network revolutions also are driving the development of low-cost sensors and (relatively) low-cost storage systems. Major "big science" efforts such as the International Geosphere-Biosphere Programme (IGBP) and the Human Genome Project involve the collection and distribution of large volumes of data

TABLE 2.2 Summary of Technical Trends Affecting Exchange of Scientific Data and Information

Trend	Description
Decreasing cost of computing and communications; "technology leapfrogging"	Following Moore's Law, the cost of computing, data storage, and communication has fallen consistently for more than 25 years. Developing countries and the newly independent states of the former Soviet Union have, in some cases, been able to acquire modern communications and computing equipment. New users have been able to avoid substantial capital expense and the burden of depreciation of that investment.
Enhanced capabilities for collecting scientific and other data	The collection power of their instruments enables major scientific enterprises such as the Human Genome Project, climate modeling, and satellite remote sensing studies to generate very large volumes of data.
Increasing exploitation of broadband networks and emerging dominance of the video data type in networks	The investment in fiber-optic cable over the past two decades is increasingly being exploited to support demanding new applications with high-capacity or real-time delivery requirements (video, medical imaging, large-scale science). The entertainment industry and new applications such as video teleconferencing, movies on demand, and interactive television have attracted substantial investment and will be the dominant factors in the development of networks in the next 10 years. Voice communication will require a minor share of telecommunications capacity.
Advent of digital wireless communications	Wireless networks are rapidly connecting the world in new ways, and at low cost. Ground-based wireless systems are creating modern infrastructure in cities that have had unreliable phone systems with inadequate capacity. Proposed satellite ventures will provide data and voice connections on a global basis.

Shifting dominance in data networks from primarily science/defense to commercial/entertainment applications	The Internet was developed to support advanced science and technology activities. Recent changes (in particular, the advent of World Wide Web browsers) have transformed the Internet into a tool for a vast array of both commercial and noncommercial applications (including shopping, entertainment, education, and general publication).
Increasing facility in collaborative work	Teams of scientists (remote from each other and often in different countries) are able to work together on a project, facilitated by high-performance communication for active, real-time interaction with each other using data and other information resources.
Increasing capabilities for language processing	Machines using natural language processing techniques are helping to organize the vast amount of information available in electronic form. New tools are providing transparent access (via rudimentary machine translation) for speakers of the world's major languages.
Increasing recognition of the importance of standards	Standards provide the means for interoperability and help to support competition and product evolution. Recognition of the role of standards (whether de facto, industry driven, or supported by formal national or international bodies) has grown, further accelerating the acceptance and applications of standards.
Growing acceptance of a need for cooperation in monitoring and controlling network activity	Mechanisms have been built into authentication systems, retrieval systems, and networks to account for specific activities of users and to support flexible billing systems. Public-key encryption technology is increasingly accepted as a means to protect data and authenticate users. This activity is being driven primarily by the needs of commercial users of the network.
Increasing use of intranets	The use of dedicated networks, particularly among private firms, is growing.

and data products. Other observational science and engineering projects[4] involving large-scale models, simulations, or sampling volumes also produce enormous quantities of data. NASA's Earth Observing System (EOS) is perhaps the best known example of a high-volume, long-term scientific observational system.[5] EOS is expected to collect a terabyte per day of satellite sensor data by the beginning of the next century.

The desire to collect, manage, and preserve scientific information always appears to exceed the financial and technical capabilities to do so, even in the wealthiest nations. Scientific communities must organize themselves better to select information for acquisition and for retention.

Advent of Digital Wireless Communications

Wireless communications received a major boost from the effort to develop mobile communications systems in the United States. Interest and investment also have been stimulated by the possibility of creating competition in local telephone service, heretofore a 100-year monopoly. Moreover, the end of the Cold War has forced aerospace companies to seek new markets for satellite technology, including direct-broadcast television and satellite-based cellular telephony. Wireless communications links are being installed worldwide, enabling mobile communication—and, for developing countries and other nations with historically weak telecommunications infrastructure and rapid growth, avoidance of much of the capital cost of a wired communication infrastructure. New competition will drive down the cost of telephony and offer new applications. Video broadcast from space or from fixed terrestrial sites may offer new ways to deliver data in interactive communications systems.

Increasing Exploitation of Broadband Networks and Capabilities for Transmission of Video Data

Commercial providers believe that new applications such as video conferencing, interactive television, and the ability to access movies on demand from a large archive will be the dominant factors in the development of networks over the next 10 years. Voice communications will require an ever smaller share of telecommunications capacity.

The widely discussed convergence of personal computers and television has been accelerated through the widespread licensing of new tools for interactive World Wide Web (WWW) applications and through emerging standards by which cable television companies can provide high-speed Internet access. Much of this activity is driven by the goal of providing interactive access to large video databases in "real time" (at least 1 megabit per second).

New, higher-bandwidth protocols such as the high-performance parallel interface (HIPPI), the first gigabit-per-second standard begun at Los Alamos National Laboratories in the early 1990s, and the MBone (a virtual multicast backbone network for delivery of audio and rudimentary real-time video across the Internet) are being developed. In the short term, however, the impact of high-bandwidth applications will be negative (especially for high-data-rate users in OECD countries), since the need for higher bandwidth has already been outpacing bandwidth improvements, both on major backbone networks and on bridges between them (see the section below titled "Specific Technical Concerns").

Shifting Dominance in Data Networks

The international public infrastructure for data communications is built around the Internet. Originally developed in the United States by the Department of Defense, the National Science Foundation, and other agencies to support scientific and technical collaboration,[6] the Internet now serves a much wider range of purposes. In recent years, it has become a high-visibility source of entertainment as well as an indispensable tool for many commercial and noncommercial applications (e.g., catalog sales, news, social interaction, dissemination of company and product information). Advertisers use the Internet to promote themselves and their wares as "high tech" and, moreover, view the current demographics of Internet users (who have disposable incomes that are typically much higher than average) as extremely favorable.

In 1995, the total number of commercial (".com") sites on the Internet grew to exceed the number of educational and government sites for the first time, and this continues to be the sector of most rapid growth. For example, the percentage of Web sites on the Internet running from the ".com" domain in the United States increased from 1.5 percent in June 1993 to 50 percent in January 1996.[7]

This trend toward commercial use of the Internet could have a significant impact on the scientific community. What has been until now a government-subsidized activity could become a significant cost factor to scientists as networks become privatized. Further, the scientific community originally played a major role in developing the technologies and standards for the Internet, but this is no longer the case. Scientific activity will have to follow (and potentially benefit from or suffer because of) the standards and pace set by others.

Increasing Technical Support for Collaborative Work

Scientists are increasingly aware of the importance of information technologies that facilitate collaborative work. The electronic messaging capabilities of operating systems used widely in the context of the ARPANET and in private, commercial messaging systems, as well as text retrieval systems such as IBM's STAIRS, System Development Corporation's ORBIT, NASA's RECON, Battelle's BASIS, and the work at Cornell University by Gerard Salton on SMART, provided much of the early technical framework for knowledge management and sharing. In recent years, electronic mail (e-mail) systems, mailing lists, and bulletin boards have enabled rapid information sharing among groups of people distributed throughout the world. Other commercially available computer-based tools and technologies have enhanced collaborative work by facilitating cooperative research involving, for example, the use of remote instruments, and electronic data publishing that speeds the dissemination of research results.[8] Indeed, the success of many complex scientific investigations now is predicated on bringing the capabilities of diverse researchers from multiple institutions together with state-of-the-art instruments. In addition to the purely technical issues raised by these requirements, however, the research agenda for creating such "collaboratories" must address fundamental psychosocial questions.[9]

Desktop video conferencing is a next logical step in the use of collaborative tools and may be as widely available within 10 years as e-mail is currently, provided that adequate bandwidth can be supplied. Users can now obtain rudimentary desktop video conferencing systems for as little as $100 using the CU-SeeMe software from Cornell University;[10] such systems provide crude service today but offer great promise. The Internet Engineering Task Force (IETF) and several universities are using the MBone to broadcast symposia and conference events worldwide.[11] Video conferencing systems based on integrated services digital network (ISDN) services and asynchronous transfer mode (ATM) are now available commercially, offering high-quality images and advanced application-sharing features.[12] "Plain old" telephone service (POTS)-based video conferencing is expected to be available with the next release of major PC operating systems.

The low cost of desktop video conferencing equipment and the ability to operate over a variety of media types will enable scientists who have access to these technologies to communicate more readily. These types of technologies can help improve the efficiency of scientific fieldwork, especially in remote areas, but only if they are supported by links with sufficiently high bandwidth. Investment in commercial products that support information sharing and workflow has accelerated as vendors recognize the importance of multiuser support to acquiring and sustaining market share.

Growing Capabilities for Natural Language Processing

Natural language processing has been an active branch of artificial intelligence for decades. Recent approaches and products have significantly improved automated document subject classification.[13] In addition, the Internet has greatly increased interest in capabilities for indexing and locating knowledge, thus contributing to the rapid growth of the text retrieval industry. Users can now gain more rapid access to a wider base of scientific information.[14] Moreover, numerous products (e.g., Fulcrum, Context, Limbex, InQuizit, Excaliber, Excite, Systran) and services (e.g., Digital Equipment's Alta Vista, Yahoo, Lycos, Dejanews, InfoSeek) are now using natural language processing capabilities to help organize information. More advanced products from the U.S. government's Tipster project are maturing for "information robot" ("knowbot") applications, such as agent-based information gathering, data overload filtering, and extracting key facts from raw text. These new tools accelerate work by reducing the volume of information that needs to be evaluated.

Slow but steady advances in machine translation are already beginning to produce acceptable levels of quality for some applications. New applications in handwriting and voice recognition as well as voice synthesis promise to bring the world's information resources within reach of many who previously had been excluded because of language differences or disability. The development of new language-processing capabilities is increasingly important as the historical dominance of English in data networks gives way to multilingual communications.

The ability to perform automated language translation, though still crude, facilitates global data and information access by helping users with native languages other than English to participate in scientific activities. Although current investment is limited to a small number of the languages most widely used for political and economic purposes (e.g., English, French, Chinese, Japanese, Spanish, Russian, German), advancing techniques in language processing and computer power will make extension to new language domains less costly and time consuming. Some databases, such as the European Dictionaire Automatique, have been developed explicitly to facilitate machine translation and semantic analysis.

Increasing Recognition of the Importance of Standards

Standards play a major role in the evolution of telecommunication networks because of the importance of interoperability of these networks, which also must provide for continuous paths for improvement without disruption of existing infrastructure. In computing, vendors put substantial effort into proprietary approaches to protect market share. But the U.S. government's championing of "open systems" and the Portable Operating System Interface for Computer Environments (POSIX) standards has allowed a new class of vendor to emerge and create entirely new market forces with many suppliers in every niche of computing. IBM's decision to make the PC an open, standard product provided another major force toward standardization in computing. Standards for products and for the representation of information have advanced rapidly over the last decade. Industry standards such as Transmission Control Protocol/Internet Protocol (TCP/IP), Simple Message Transfer Protocol (SMTP), Simple Network Management Protocol (SNMP), X.400, Standard Generalized Mark-up Language/HyperText Mark-up Language (SGML/HTML), and easy-to-use browser products such as Netscape and Mosaic were necessary for the rapid expansion of the Internet. Companies still use proprietary approaches to gain short-term market advantages, often with the hope that their products will become the standard (e.g., Microsoft's OLE). Sun's Java language is an interesting example of a company-sponsored effort that is becoming a standard through rapid expansion of licensing agreements.

The marketplace today often converges rapidly on one or a few standards, the standard for a high-density CD/ROM (and, more recently, digital versatile disks) being an excellent example.[15] The music and entertainment community realized that competing standards would risk an expensive competitive battle. Other examples include, among many others, the widespread application of HTML.

Technical standards increase competition and product availability, while reducing price. The downside is that standards themselves evolve and can contribute to a kind of industry-driven obsolescence. Also, when multiple standards apply in the same area, buyers are forced to try to choose prospective winners and losers (recall the battle for consumer support of the Beta and VHS standards).

Within the scientific disciplines, there is increased attention to system interoperability in terms of both data and software. In the astronomy community, for example, the interchange of data has become fairly simple because of effective coordination in the United States and internationally. Radio astronomers developed a voluntary standard format for data interchange (the flexible image transport system; FITS) that was widely adopted in the astronomy community during the 1980s. This standard is maintained by an international committee, with support from several organizations, including NASA. There are related standard formats for planetary data, as well as a trend toward the development and adoption of a few comprehensive data analysis systems that could be used with a variety of types of astronomical data from different observatories and instruments and different subdisciplines. Sharing of analysis software and commercially developed computing tools among the different systems is encouraged.

Of course, the need for standards for effective data exchange is not confined to telecommunications, computer languages, and storage media. Even within a narrow discipline or subdiscipline, true data exchange with proper interpretation of numbers, symbols, words, and graphics depends on standards for data structures, database management systems, and even terminology.

Cooperation in Monitoring and Controlling of Network Activity

The rapid growth in networks over the last 15 years has led to the need for appropriate levels of cooperative monitoring and control. Initial ad hoc activity in developing protocols such as SNMP has given way to more elaborate standards and tools today. Authentication systems, retrieval systems, and networks can now account for specific activities of users and can support flexible billing systems. Public-key encryption technology is increasingly accepted as a means of protecting data and authenticating users. Such developments are being driven by needs associated with the network as a market place.

Version 6 of the Internet Protocol (developed by IETF and often referred to as the Internet Protocol Next Generation)[16] includes the necessary technical components to support security, authentication, delivery of multimedia, and the continuing growth of the Internet. Several implementations of the proposed standard are now available.[17] As with telephones, circuit and switch technology readily supports network control and accounting during connection creation and breakdown. The ATM forum[18] is leading much of the work in this area. These initiatives will help applications developers to implement the functions necessary to make networks commercially viable.

The functionality provided by fine-grain network control will support additional penetration of network equipment and services into the market, which should in turn drive down equipment costs and provide an infrastructure to support community services such as network access to the public libraries. The technology also can be used to protect proprietary information and allow publishers and others to make better use of the Internet.

Increasing Use of Intranets

Tools for searching and creating HTML pages, internetworking hardware (e.g., routers), and strong dependency on electronic mail have supported the rapid growth within corporations of private networks known as intranets. These networks provide reliable service, high-performance access, and information protection not afforded by the public Internet. Today, sales are brisk for intranet products such as browsers, servers, and search engines for internal corporate applications. A high-profile example is the Hewlett Packard intranet, which links more than 110,000 PCs and workstations and transfers over 5 terabytes of data per month. Hewlett Packard also supports public bulletin boards with company or product information that dispense over 15 terabytes of data per month.[19]

The interest in intranets is also evident in initiatives to support priority research and education needs. Prominent among these is the Internet II project, which initially is connecting approximately 100 universities over a private, reserved backbone with 622-Mbps links.[20] This type of network could be used more broadly by the scientific community and extended to reach international partners to solve specific needs for bandwidth and for real-time control.

SPECIFIC TECHNICAL CONCERNS

Table 2.3 summarizes six major technical barriers to the international transfer of scientific data and information within the context of the trends discussed above. These are cases in which the trend, while generally favorable, produces a negative consequence or side effect.

Internet Congestion Becoming a Serious Problem[21]

The scientific and technical community once dominated the Internet and

TABLE 2.3   Summary of Major Technical Barriers to International Transfer of Scientific Data and Information

Concern	Impact
Internet congestion is becoming a serious problem.	Scientific activities are disrupted through lack of control of network capacity. High-bandwidth applications are impeded or blocked, and urgent communications are slowed.
Description and indexing of data are inadequate to support their use by others.	Data must be transformed, recomputed. Data cannot be located, and there is potential for error. Cost and delay in performing scientific work are generally increased.
Electronic storage media have limited life spans.	Data are lost or rendered unusable in the absence of long-term commitments for transferring them to new media on a regularly scheduled basis.
Tools for authentication and privacy are immature. Data and networks are vulnerable.	Valuable assets (data and infrastructure) could be lost or corrupted; intellectual integrity could be compromised. If tools depend on restricted (for export) encryption technology, barriers to free, protected exchange of information could emerge. Encryption technology for digital identification, authentication, and privacy safeguards might remain inconsistent from country to country, limiting information commerce. Alternatively, data would remain unprotected.
Scientific requirements for computer technology could be left unsatisfied by priority support for other larger market needs (e.g., entertainment and business).	Increased expense to scientists for equipment specifically tailored to research needs (e.g., supercomputers). There is the risk that specific requirements will not be met.
Few international networks support real-time data.	Lack of support for conferencing and collaborative work, large file transfer, or shared scientific infrastructure.

collectively dictated its priorities and use. Today, the network is available for a wide range of activities, which sometimes significantly reduce or block access by scientists. Even scientific events, such as the Shoemaker-Levy Comet's crash into Jupiter, have caused such high lay public interest and associated high-volume transfer of images that scientific access to research resources has been impeded or blocked. Requirements for carrying out scientific research, the results of which are often at the service of the world community on an urgent basis, now are not being served with responsive mechanisms that give them priority.

Generally, the rate of international exchange of scientific information has risen steadily as the means to carry out such exchange have improved. Now scientists on different continents commonly share ideas and data daily, or even hourly. A researcher in one country can remotely connect to a computer in a different country to perform calculations and data analyses and sometimes complete experiments. When linking to a remote computer, one expects—or at least hopes—to transfer information between the local and remote computers at approximately the rate at which data move in the local computer alone. As the Internet developed in its first decade, it usually operated in this way.

But as the Internet's popularity and use have grown, more people are expecting to get instantaneous service for all their activities. These include using the Web as well as linking to a distant site to extract data of all sorts (from accumulated electronic mail to numerical tables to animations) or to run programs and obtain the results remotely. As the speed and efficiency of computers have improved, scientists using them have generated and have expected to obtain ever more complex kinds of data. Images, particularly animations, require extremely large data sets if they are to be transmitted electronically.[22] In fact, the burgeoning interest in images and video animation has led to a dramatic increase in the amount of information people want to transmit over the Internet, especially as more people use the Web and as the Web serves more commercial and entertainment functions.

This explosion of use has strained the carrying capacity of the Internet, particularly for many of the most heavily used intercontinental links.[23] Direct trans-Atlantic links between the United States and Germany, for example, functioned virtually as efficiently as local area networks until sometime in 1993 or 1994. Then delays began to occur at about 12:00 or 1:00 p.m. GMT when users on both sides of the Atlantic are active. The delays have grown ever longer, as has the time period during which delays occur, so that now, from about 8:00 a.m. until midnight GMT, delays can be so lengthy that the user is "timed off" by the system in the middle of a (delayed) transaction—not once, but many times. This problem apparently can occur even with data exchanges such as e-mail, which computers transmit whenever the lines are open. In nations with very limited network or gateway facilities, e-mail may take a day or more to be transmitted to or from some countries. Since the inception of this study network congestion and delays have become severe.

If the Internet were to become saturated, it would be rendered ineffective as a means of transmitting scientific information directly. Although there are several satellite and undersea fiber-optic cable systems currently being developed that may be expected to supply sufficient transmission capacity worldwide, some near-term remedies will be necessary to ensure that scientists and others with professional needs will be able to continue using the Internet with at least moderate efficiency, until the new systems become operational.

Inadequate Description and Indexing of Data

Responses to the committee's "Inquiry to Interested Parties" (see Appendix D) revealed the lack of common representations for data to be the primary *technical* challenge for international scientific exchange. Interdisciplinary, collaborative work builds on shared understanding and agreement on terminology. Standards for representation of data, including units and formats, as well as description (metadata), are vital. Effective directories and navigation tools are needed to help scientists locate relevant information. Shared understanding of the operation of algorithms is important to the application of information. This understanding must also evolve coherently over time as algorithms, standards, and collection instruments are developed. For all scientists, the lack of shared understanding can lead to duplicated effort, additional work to "normalize" data, or limited capability to integrate research results. In extreme cases, information collectors duplicate each other's information, because they have no a priori agreement on effective data representations. Problems of data compatibility and integration, even within the United States alone, were reviewed in some detail in a 1987 CODATA conference from three different perspectives: government, geography, and technology.[24]

Rapid Obsolescence of Electronic Storage Media

The media on which scientific data are stored are vulnerable to decay and obsolescence. The standard lifetime of a particular disk or tape appears to be less than a decade; the data stored on these media must be copied or refreshed at regular intervals. A recent National Research Council study[25] discussed the effects and implications of long-term commitments in scientific data management, with respect to both selection of data for long-term retention and media obsolescence. Further (and paradoxically), data collected before the advent of computers and stored on "archival media" (paper) must be put into electronic form to be used widely and effectively today. Such data can add enormous value to research efforts, particularly for studies examining long-term trends, but are costly to transform.[26]

Valuable records may fail to be transferred to new media or transformed to electronic form because of a lack of resources (funds or appropriate equipment) or lack of motivation. Scientists without long-term support commitments will face the discouraging fate of losing precious data assets over the long term. With the extraordinary volumes of data being collected, transferring data to new media and managing high-value data sets for active use will increasingly challenge the scientific community, particularly since the time frames for rescuing old, deteriorating data are frequently quite short. Examples include scientific publications printed on high-acid paper and data sets stored on magnetic tapes that are crumbling, some after only a score of years.

Vulnerability of Electronic Data Networks

Some scientific data must be treated with special care to ensure their dissemination only within a prescribed community (e.g., to protect the privacy of individuals, to allow for verification of results, or to maintain the proprietary advantage of a private enterprise). Today, tools for authentication and for protecting privacy of data are difficult to use, do not follow widespread standards, and, in some cases, involve encryption technologies that cannot be universally distributed. Such tools, however, can help researchers to maintain control over the research environment. Because effective use of authentication and privacy measures involves the collaborative effort of numerous scientists or institutions, "top-down" leadership in standards setting across the scientific community may help speed the acceptance and use of emerging tools designed primarily for electronic commerce on the Internet.

Beyond the basic issue of protecting information privacy and integrity, the scientific community must prepare itself for disruptions of basic network and computing infrastructure. The complexity of software and networks, the large number of users, the dynamic changes in staff, and the relative sophistication of programmers worldwide leave networks vulnerable to attack and catastrophic accidents. We already have experienced large-scale disruption of the Internet and telephone systems. International scientific data collection and dissemination activities are similarly vulnerable to both intentional and unintentional disruptions. A proper balance thus needs to be maintained between open, but vulnerable, access and secure, but not overly rigid, control.

Scientific Requirements for Computer Hardware Potentially Unmet

The demands of entertainment (in particular, interactive multimedia, animation creation, and delivery of large numbers of simultaneous video channels) are driving the frontiers of computer and communications technology. It is possible that computers and networks will be optimized for entertainment applications, making it difficult for some scientific endeavors to have access to cost-effective computer power. For example, a computer optimized for video streaming might not be suitable for running a large chemical model or ocean simulation. In the past, scientific applications—and funding for the advanced computers to support them—have driven computer design. Advances such as floating point and vector accelerators, massively parallel computers, gigabit networks, large-volume storage media, and visualization software were developed because of scientific needs. Continued funding on an international basis for research leading to these kinds of advances is necessary if vendors are to respond to the technical needs of science. The goal is to incorporate advanced features for scientists within commercially available products.

Lack of Sufficient International Real-Time Data Networks

Scientists increasingly need real-time communications capabilities for collaborative scientific activities and optimal use of major experimental and observational facilities. Advanced networking and computing services can make large file transfers practical and provide for remote access to (and control of) large-scale scientific and medical facilities. However, even with the growth of new telecommunications capacity, the availability of high-bandwidth circuits to data acquisition and analysis sites and to the desktop will continue to lag behind demand. Also, the Internet protocols that are now in wide use do not effectively support time-synchronized activity. Circuits to the developing countries will be limited to the relatively low speeds of voice circuits until new (submarine and extension fiber) cable connects all corners of the world and affordable capacity becomes available. Wireless communication systems will operate at lower speed than comparable wired services for the most cost-effective use of spectrum. In some countries the existence of outmoded government-operated facilities could impede the development of new, high-speed or alternative-capacity links because government-run post, telephone, and telegraph ministries (PTTs) are used to subsidize nontelecommunication governmental activities and maintain monopolistic control over access and use.

DATA ACCESS ISSUES IN DEVELOPING COUNTRIES

Although at first sight, the gap between the "haves" and "have nots" for access to scientific data and information seems to widen each day, the long-term outlook for such access in developing countries is far better than it was before the advent of electronic communications, primarily because the cost of the technology continues to decline even as the capabilities improve. It is potentially more cost-effective to buy computers, networks, and mirror sites of the libraries and data centers in developed countries than to try to maintain autonomous libraries with up-to-date collections of books, journals, and data compilations.

As scientists in developing nations obtain computers with connections to networks linking them to international collections of scientific information, they greatly increase their research capabilities. Low-cost computers and modern software approaches are available to help developing countries "leapfrog" multiple generations of equipment and approaches. Satellite ventures are planned to provide worldwide access for short messages, voice telephony, and broadband digital links.[27] Direct-broadcast television also is having an impact by offering hundreds of channels of high-quality video to a growing percentage of the world's population.

Many of the developing countries may soon be expected to be the beneficiaries of a broadly distributed modern telecommunications infrastructure. Scientific efforts in these countries will be supported, even in remote areas. For example, the planned Teledesic wideband satellite communications system has pledged to give excess capacity to developing countries for a variety of uses, including applications in education, science, and medicine.[28] Direct broadcast television could serve to raise education levels. The anticipated low cost of desktop video conferencing equipment and the ability to communicate with multimedia functions, as described above, can enable scientists and others in the less developed countries to participate more fully in global scientific research, subject to the availability of high-bandwidth transmission capabilities and the mitigation of local cost barriers. Hardware and software for electronic communication in the sciences therefore offer particularly high leverage for return on investment in foreign aid to developing nations.

Unfortunately, many of these technologies are not yet widely available in the most developed nations, much less in the developing countries. For example, roughly half the nations now served by some form of Internet connection have access only to electronic mail.[29] Even in those countries with at least one full-service Internet node, the proportion of users actually accessing all services is low. In addition, as noted throughout this report, full Internet service does not automatically mean full access to information.

For example, the University of Chile has high-speed Internet service. Two components of the U.S. National Institutes of Health—the National Library of Medicine and the National Cancer Institute—provide Chilean researchers with excellent Internet access to abstracts from journals through outreach programs. Although requests for search results are answered quickly, the journals themselves usually are not in the Chilean libraries. Thus the researchers frequently must wait for months to receive the full-text reprints of published papers. This situation is improving, however, and the system will become truly effective with full-text on-line access to journals. Nevertheless, most scientists in less developed nations now have little or no access to the Internet and the World Wide Web and still must depend on inadequate library facilities for full-text access to scientific data and literature.

There are various other reasons for low Internet usage in developing nations, aside from the lack of infrastructure. These include internal institutional policies stipulating the placement of computers in administrative offices rather than in laboratories, governmental restrictions on the free flow of information, poor-quality telecommunications systems, and, most commonly, lack of funds for use of whatever communications infrastructure does exist. The costs of telephone services, for example, generally bear an inverse relationship to the per capita income of a country. International calls that cost $1.00 when originating in the United States frequently cost many times that when originating in a less developed country or a country where the tariffs are a general source of revenue for the government. For example, a call from Nairobi or Moscow to Washington, D.C., can cost seven times as much as the same call originating in Washington, with the higher costs often being borne by those least able to afford them.

Various strategies can help mitigate these differences or eliminate the end user's cost entirely. The Internet itself, by institutionalizing communications facilities, can make the costs transparent to the end user. However, persons in developing nations often must limit their Internet use or drop their subscriptions to list servers because of cost.[30] One biologist in Indonesia who dropped his subscription observed that the communication costs per month were considerably more than his salary and that his institution was passing these costs on to the end users. A number of Kenyan scientists have obtained calling cards from U.S. providers and are directly dialing the United States. The billing is to their Kenyan address. People in other countries also have adopted this strategy.

One current approach to reducing communication costs in African countries is the use of the message-forwarding Fidonet system, a low-cost network of individual computerized bulletin board services that uses regular dial-up telephone lines and high-speed modems to transfer electronic messages.[31] Although most of Africa currently lacks direct TCP/IP Internet and WWW connections,[32] individuals can send and receive electronic mail via the Fidonet service of the Association for Progressive Communications, a U.S. nongovernmental organization dedicated to bringing low-cost communications to developing nations throughout the world.[33]

On other continents, the situation is somewhat better with respect to direct Internet access. However, even where Internet connections do exist, access still tends to be spotty in all but the most prestigious or centrally located institutions.

For scientists in developing countries, another difficulty is competition for access to large remote data sets, which is made even more difficult by the increasing volume of data, particularly from new observational sensors. In addition, given the vast amount of data being collected, small data sets that they might contribute may be viewed as less important, limiting the ways in which researchers in the developing countries can participate in the scientific community. One result of such disparities is the perception by some scientists in developing countries that the OECD countries take information but seldom return it on an equitable basis.

Currently, developing countries severely lag the OECD countries in bandwidth for emerging applications.[34] If the majority of communication in developing countries is wireless, end users may not be able to take advantage of the more bandwidth-intensive applications. Moreover, as noted above, problems arise even after advanced communication capabilities are installed. Transoceanic and intercontinental communication and exchange of scientific information must compete with all the other electronic traffic—increasingly business and entertainment. Unless bandwidth is improved, the "information superhighway" becomes the electronic equivalent of many urban highways during rush hour. Furthermore, in many of the developing nations, the decreasing costs and increasing bandwidths that might be available generally are not passed on to the scientific end user by the government communications monopolies.

RECOMMENDATIONS ON ISSUES IN INFORMATION TECHNOLOGY

Based on the areas of concern discussed above, the committee makes the following recommendations for improving technical support for the international flow of scientific data and information.

1. The principal scientific societies and the Internet Engineering Task Force (IETF) should begin a long-term planning effort to assess the carrying capacity and distribution capability of the Internet, using projections of storage and transmission capacity and of demand and taking into account the next generation of Internet protocols. Scientific societies should encourage their publication committees to maintain contact with the IETF and keep their members abreast of advances in technologies useful for scientific information management. One option that science societies and government science agencies should evaluate is the creation of dedicated international science networks, such as the Internet II now being developed.

2. To improve the technical organization and management of scientific data, the scientific community, through the government science agencies, professional societies, and the actions of individual scientists, should do the following:
  a. Work with the information and computer science communities to increase their involvement in scientific information management;
  b. Support computer science research in database technology, particularly to strengthen standards for self-describing data representations, efficient storage of large data sets, and integration of standards for configuration management;
  c. Improve science education and the reward system in the area of scientific data management. Provide incentives and recognition for papers dealing with data representation standards, archiving strategies, data set creation, data evaluation, data directories, and service to users.
  d. Encourage the funding of data compilation and evaluation projects, and of data rescue efforts for important data sets in transient or obsolete forms, especially by scientists in developing countries where substantial cadres of highly educated scientists exist who are underemployed and relatively inexpensive to support.

3. U.S. government science agencies, working with their counterparts in other nations, should improve data authentication and apply security safeguards more vigorously. They should implement the means to protect data, including safe storage of data copies, and support policies that make it easier to exchange encryption technology.[35]

Government science agencies also should continue funding for research and development in information technologies that are important to the pursuit of science. Examples include high-performance computing and communications, advanced database technology, higher-density storage media, and basic research in microelectronics.

4. A consortium of intergovernmental and nongovernmental organizations concerned with the international exchange of scientific data and information—including the International Telecommunications Union, the World Bank, the U.N. Environment Programme, U.N. Industrial Development Organization, U.N. Commission on Economic Development, and other Specialized Agencies of the United Nations, as well as the International Council of Scientific Unions—should mount a global effort to reduce telecommunications tariffs to scientists in developing countries through differential pricing or direct subsidy. This reduction in tariffs would have to be coupled with more timely access to new telephone lines in some countries. The result would be increasing rates of scientific data transfer in the developing countries and a significant improvement in their research capabilities and economic development.

5. Foreign aid to developing countries in the form of computers, computer networks, and associated software, coupled with the training and resources necessary to operate and maintain those technologies, should be given high priority, on the basis of the potential for long-term socioeconomic returns. The communication systems must have adequate carrying capacity to meet growing demand.

NOTES

1. Moore's Law, named for Intel founder Gordon Moore, predicts that the density of microprocessors will double every 18 months, thus halving the price. The by-product of this long-lived phenomenon has been the doubling of processor speed in the same 18-month period. Moore's "Law" is in fact a representation of the speed of change of the microelectronics industry over the last 20 years. It is expected to continue to apply to technology change for at least the next 5 to 10 years. See Ashley Dunn (1996), "The Demise of Moore's Law Signals the Digital Frontier's End," *New York Times*, August 14, at <http://www.nytimes./com/library/cyber/surf/0814surf. html>. See also <http://www-us-east.intel.com/product/tech briefs/man_bnch.html>.
2. See Brian Grimes (1995), "Modeling and Forecasting the Information Sciences," Knowledge Science Institute, University of Calgary, at <http://ksi.cpsc.ucalgary.ca/articles/BRETAM/InfSci/> for a discussion of exponential change in the performance of these technologies and its impact.
3. One terabyte is 10^{12} bytes, or 1,000 gigabytes. It is roughly the equivalent of 40,000 4-drawer files holding 500 million pages of paper documents.
4. Oryx Energy Co. estimates that in petroleum prospecting a three-dimensional seismic survey of a 3-square-mile block in the Gulf of Mexico involved hundreds of gigabits of data, requiring three months of supercomputer time to digest. See J. Dubashi (1990), "Images and Imaginations," *Financial World*, 24 (July):8.

5. See NASA's EOS Project Science home page at <http://eospso.gsfc.nasa.gov/> for additional information and related sites.
6. For a description of the origin of the Internet, see Vinton Cerf (1996), "Computer Networking: Global Infrastructure Drive for the 21st Century," *On the Internet,* 1(5):18-27.
7. See Matthew Gray of the Massachusetts Institute of Technology (1996), "Web Growth Summary," at <http://www.mit.edu:8001/people/mkgray/net/web-growth-summary.html>.
8. National Research Council (1993), *National Collaboratories: Applying Information Technology for Scientific Research,* Computer Science and Telecommunications Board, National Academy Press, Washington, D.C.
9. Richard T. Kouzes, James D. Myers, and William A. Wulf (1996), "Collaboratories: Doing Science on the Internet," *IEEE Computer,* 29(8):40-46.
10. See the Cornell University CU-See Me Welcome Page at <http://cu-seeme.cornell.edu>.
11. See the MBone Information Web home page, sponsored by ICAST Communication, Inc., at <http://www.best.com/~prince/techinfo/mbone.html>.
12. See the Bits Scout Software home page at <http://www.bitscout.com/>.
13. Kenneth W. Church and Lisa F. Rau (1995), "Commercial Applications of Natural Language Processing," *Communications of the ACM,* November:71-79. See also the home page of the American Society for Information Science at <http://www.asis.org>.
14. A recent prototype is the Net Advance of Physics, a master hierarchical index of on-line review articles in physics. It is hoped that this free service eventually will be expanded to include the entire content of the physics e-print archives (see *Science* 272 (1996):15-16).
15. A technology going beyond CD-ROM is HD-ROM (high density-read only memory), originally developed at Los Alamos National Laboratory, which gets much greater storage density than conventional CD-ROMs at a fraction of the cost. It uses an ion beam to etch pins of stainless steel, iridium, or other similarly long-lasting materials. The etching is done in a vacuum, which allows the high densities, but the reading can be done in air. It is now in the process of commercialization. The DVD standard is being supported by Sony and Toshiba. See <http://www.islandtel.com/newsbytes/headline/dvddisputeen/dsincompro_350.html>.
16. See the IP Next Generation (IPng) home page, developed by Robert Hinden of Ipsilon Networks, Inc., at <http://playground.sun.com/pub/ipng/html/ipng-main.html>.
17. See the implementations IPng home page at <http://playground.sun.com/pub/ipng/html/ipng-implementations.html>.
18. See the ATM Forum Communications home page developed by Raj Jain's group at Ohio State University in Columbus, Ohio, at <http://www.cis.ohio-state.edu/~jain/atmforum.htm>.
19. Caryn Gillooly (1996), "Blazing a Trail in Intranet Usage," *Information Week* (September 9): 98-102; additional information on Intranet products can be found on the *Internet Design* magazine home page at <http://www.innergy.com>.
20. See White House Press Release, "Background on Clinton-Gore Administration's Next-Generation Internet Initiative," Office of the Press Secretary, October 10, 1996, Washington, D.C.
21. The economic aspects of Internet congestion are discussed in Chapter 4 in the section titled "Electronic Access and Internet Congestion."
22. Text is far less demanding; for example, the 20-volume *Oxford English Dictionary* is available now as a single CD-ROM. However, storing a full-length movie on a single CD-ROM will be feasible only when the new, high-density CDs, with much greater storage capacity than the current variety, become available.
23. See <http://www.nlanr.net/ISMA/Report/#background> for the results of the NSF-sponsored workshop on Internet-statistics measurement and analysis; also the Corporation for National Research Initiatives' home page at <http://www.cnri.reston.va.us>, and "The Interminablenet," *The Economist,* February 3, 1996, pp. 70-71.
24. L.J. Allison, and R.J. Olson, eds. (1988), *Piecing the Puzzle Together: A Conference on Integrating Data for Decision Making,* U.S. National Committee for CODATA, Integrated Data

Users Workshop, National Governors Council, published by the National Governors Association, Washington, DC: 264 pp. plus Appendices. For a more recent study regarding the barriers and issues inherent in data integration efforts, see National Research Council (1995), *Finding the Forest in the Trees: The Challenge of Combining Diverse Environmental Data,* National Academy Press, Washington, D.C.

25. National Research Council (1995), *Preserving Scientific Data on Our Physical Universe: A New Strategy for Archiving the Nation's Scientific Information Resources,* National Academy Press, Washington, D.C.

26. W. Grattidge, J.H. Westbrook, C. Brown, and W.B. Novinger (1987), "A Versatile Data Capture System for Archival Graphics and Text," *Computer Handling and Dissemination of Data,* ed. P.S. Glaeser, Elsevier Science Publishers B.V. (North Holland), CODATA; W. Grattidge, W.B. Lund, and J.H. Westbrook (1990), "A Data Capture System for Printed Tabular Data," *Proc. 11th Int'l CODATA Conf.: Scientific and Technical Data in a New Era,* P.S. Glaeser, ed., Hemisphere Press, Karlsruhe, FRG, 302-306; W. Grattidge, W.B. Lund, and J.H. Westbrook (1992), "Problems of Interpretation and Representation in the Computerization of a Printed Reference Work on Materials Data," *Computerization and Networking of Materials Databases,* Vol. 3, ASTM STP 1140, Thomas L. Barry and Keith W. Reynard, eds., American Society for Testing and Materials, Philadelphia.

27. See the Norwegian University of Science and Technology Department of Computer Systems home page at <http://www.idt.unit.no/>.

28. Patrick Seitz (1995), "Firms Battle for Spectrum," *Space News,* November 27:1.

29. International Internet connectivity levels improve constantly and cannot be generalized about. For a continuously updated report, see the Internet Connecting Chart, copyright by Larry Landweber 1995, at <http://www.infopro.spb.su:8000/info/internet/table_eng.html>. See also the International E-mail Accessibility home page compiled by Oliver M.J. Crepin-Leblond at <http://www.ee.ic.ac.uk/misc/country-codes.html>; e-mail access information is based on International Organization for Standardization (ISO) standard 3166 names. For a survey of International Internet and K-12 Connectivity done by the NASA Science Internet Program, compiled by Antony Villasenor, see <http://nic.nasa.gov/ni/survey/survey.html>.

30. See, generally, the series of articles in "Eye on Emerging Nations" in *On the Internet,* the Internet Society, Reston, Va.

31. National Research Council (1996), *Bridge Builders: African Experiences with Information and Communication Technology,* National Academy Press, Washington, D.C. See also the African Academy of Sciences/American Association for the Advancement of Science (1993), *Electronic Networking in Africa,* AAAS, Washington, D.C.

32. See the U.S. Agency for International Development's Africa Link home page at <http://www.info.usaid.gov/alnk/connect/conmap.html> for a detailed description of Internet and other types of network connectivity on the African continent.

33. See <http://www.apc.org/about.html> for more information about the activities of the Association for Progressive Communications in Africa and other regions of the world.

34. See Trudy Bell, John Adam, and Sue Lowe (1996), "Communications," *IEEE Spectrum* (January):40.

35. See National Research Council (1996), *Cryptography's Role in Securing the Information Society,* Computer Science and Telecommunications Board, National Academy Press, Washington, D.C.

3

Scientific Issues in the International Exchange of Data in the Natural Sciences

Science is the process and the product of discovering the cumulative body of knowledge and understanding through which we humans comprehend the tangible universe. Its cumulative nature is a key to the uniqueness of the knowledge gained in the natural sciences. This knowledge is sometimes reorganized at a profound conceptual level when a field undergoes a shift of paradigm—for example, the change from the caloric fluid to the kinetic theory of heat or from the continuum of classical mechanics to the discreteness and duality of quantum mechanics. Yet the facts of science and the links among them remain; we may change the way we interpret those links, but the body of scientific data continues to accumulate.

Data in science are like bricks, and the theoretical concepts are the mortar that connects them to give a subject its structure. Each new bit of data plays a part: it may be uncovered in efforts to test a hypothesis, estimated from previous information, or collected in observations, experiments, or computations. As an observed or measured new piece of information, it becomes part of our base of knowledge, to share, interpret, and reconcile with the data already in hand. Scientists ask, "Are these new data consistent with what we already know? Are they just what we might have expected, or do they require us to question the results, to repeat the experiment, or to find a new interpretation that accounts for why the data are what they are?" When the scientific community resolves these questions, the new data become part of the foundation on which the next conjectures and experimental plans build. At this stage, also, researchers begin to consider the implications of the new data, both to strengthen and extend basic understanding in the natural sciences and to seek applications that may bring benefits to society and progress in bettering the human condition. Throughout this process, scientific data are the cumulative substance on which all of science builds.

Data in science are universal—they have the same validity for scientists everywhere. The atomic mass of iron, the structure of DNA, and the amount of rainfall in Manaus in 1972 are facts independent of the political views of their user, the time at which we determine them (apart from the evolving, improving accuracy of the determinations), or the user's location. Their utility depends on the precision and accuracy with which they are determined and the units we use to express them. A DNA sequence or a nuclear cross section can be as important to a researcher in Novosibirsk as it is to another in Pasadena. Consequently, except in situations involving national security, the protection of individual privacy,[1] or proprietary rights, scientists have developed an ethic of full and open exchange of data, within and across national boundaries. Although infringements occasionally do occur, they typically generate community disapproval. Full and open exchange of information is a fundamental tenet of basic science that scientists regard as essential to optimizing their own work and that of their colleagues, as well as to enabling the advance of science overall.[2]

Traditionally, scientific data were compilations in lists, tables, and books—essentially all on paper—which circulated like all other scholarly information, through personal exchanges, subscriptions, and libraries. Today, electronic handling of scientific information is becoming the norm. With this evolution has come a dramatic increase in the international scope of scientific cooperation and exchange of information. While basic science has always been largely a collaborative activity that readily crossed national boundaries, electronic communication has made this cooperation much more informal, intimate, instantaneous, and continuous than ever before. Consequently, scientific data now may flow between scientists in different parts of the world as if they were across the street.

Scientists have been, to a large extent, the creators of the means and the environment for the ethical code governing the open exchange of their data. This is as true in the evolving electronic environment as it has been in the past. Now, however, interests outside the scientific community are exerting forces on that environment that could severely restrict this open exchange. Scientists believe that restrictions on data access will slow the progress of science and significantly diminish the potential benefits that science renders to society.

An important consideration in any discussion of exchange of scientific data concerns the "market" in which scientists participate, and particularly what its "goods" and "return" are. Scientists in academia and government are motivated overwhelmingly by the desire to generate ideas that influence the course of science. They want their papers to be read, so much so that they regularly pay page charges to have those papers published. Traditional concepts of copyright, protection of intellectual property, and financial return to the creator of a written work may apply to a scientist who writes a textbook, but become irrelevant to the researcher publishing a paper in a scientific journal. Publishers of such journals, sometimes including professional societies, adhere to traditional motivations for protection of intellectual property and copyright. Scientists are usually delighted when someone wants to photocopy their articles; their publishers are sometimes aghast at the same photocopying act. This tension is often overlooked in considerations of adapting to electronic exchange of scientific information. It becomes especially important when one tries to bring economic and legal thinking to bear on the management of scientific data, and on the behavior and the system of values of scientists. (For more detailed discussion on these issues, see Chapters 4 and 5.)

This report focuses primarily on international access to scientific data for basic research purposes. Nevertheless, in some disciplines, such as meteorology, a significant part of the data is generated to serve the general public by making possible severe weather and flood warnings and associated weather prediction. In formulating policies for international data exchange, the need for data for these applications also must be taken into account. In this chapter the committee broadly characterizes types of scientific data and their use in the laboratory physical sciences, astronomy and space sciences, Earth sciences, and biological sciences; outlines some of the major data trends, opportunities, and challenges in the natural sciences; discusses selected discipline-specific issues; and describes problems of access to data in less developed countries. The chapter concludes with the committee's recommendations for steps to improve access to data in the natural sciences worldwide.

TYPES OF DATA AND THEIR USE IN DIFFERENT DISCIPLINES

There are several ways to characterize scientific data: among others, by form, whether numerical, symbolic, still image, animation, or some other; by the way they were generated or gathered, that is, from experiment, observation, or simulation; by level of quality; by the size or form of the databases that contain them; by the nature of the support for their generation or distribution, that is, public or private, national or international; and, of course, by subject. Perhaps the most obvious differentiation is according to the degree of refinement of data along the path from collection to publication. Several linked levels of data can be distinguished in this hierarchy, beginning with initially collected experimental or observational data.

In the laboratory sciences today, data at this first level are rarely raw readings or counts. Sophisticated means for gathering and manipulating such information have softened the concept of "primary" data. The computer mediating an experiment is likely to extract from several measurements some average of the total signal minus the background noise. Frequently, the first data an experimenter sees appear as a curve or a set of points that represents addition, subtraction, and averaging of several kinds of measurements, all collected and manipulated electronically. While some of these data may be published as tables, most data at this level have limited distribution. They are useful when shared among the participants in a large collaboration, for example, in a high-energy physics experiment. International distribution of data of this kind is normal practice, particularly among collaborating scientists.

The second major level of data in the laboratory sciences is usually published scientific results based on collected data, sometimes including the data and sometimes only providing a pathway by which the data can be obtained. Evaluated data files, the next level in the hierarchy, are compilations of data from several sources created when an "evaluator" has worked to obtain the "best" values of the tabulated quantities. Such files are often broadly disseminated, sometimes in journals established for that purpose, such as the *Journal of Physical and Chemical Reference Data*; increasingly, these files will be available electronically and, with hypertext, will be linked, so that anyone reading a manuscript will have ready access to on-line data files on which published results are based, just by clicking on the relevant figure or text. When data are structured or compiled in an organized manner, whether in raw form or after thorough evaluation or processing, they become a database.

In the observational sciences, scientific research leads to the generation of data that can be processed and interpreted at different levels of complexity.[3] Typically, each level of processing adds value to the original, raw data by summarizing the original product, synthesizing a new product, or providing an interpretation of the original data. The processing of data leads to an inherent paradox that may not be readily apparent. The original unprocessed, or minimally processed, data are usually the most difficult to understand or use by anyone other than the expert primary user. With every successive level of processing, the data tend to become more understandable and better documented for the nonexpert user. One might therefore assume that it is the most highly processed data that have the greatest value for long-term preservation and international exchange, as in the case of the laboratory sciences, because they are more easily understood by a broader spectrum of potential users. In fact, just the opposite is usually the case for observational data, because it is only with the original unprocessed data that it will be possible to recreate all other levels of processed data and data products. To do so, however, requires preservation of the necessary information about the processing steps and ancillary data.

Another important way to characterize scientific data in general is by quality, as indicated by their degree of acceptance in the scientific community. "Prepublication" data bear no certification whatsoever. Such data would, for example, be considered by most scientists to be inappropriate as legal evidence. Data accepted for publication in a refereed journal carry a certification that they, and the text that accompanies them, contain no obvious error and are admissible topics of scientific discourse. Published data, however, are often challenged, and occasionally the data or their interpretations prove erroneous. When they have been thoroughly validated, data become dogma. Values of natural constants, to some number of decimal places, are firmly established in this way. Steps toward confirmation of the soccer-ball structure of the molecule C^{60} illustrate this progression in acceptance and endorsement of data and their interpretation. At first it was conjectured, prior to publication; then the proposed structure was published and shown to be consistent with other evidence then available. When a method appeared for preparing the substance in macroscopic quantities, new experiments, notably x-ray diffraction, nuclear magnetic resonance, and infrared spectroscopy, gave unassailable proof that the molecule is indeed shaped like a soccer ball. Since then, nobody would think of questioning that structure.

Particular uses of data and characteristics of disciplines in the natural sciences influence needs for and conditions affecting global access to information in those areas of research. Examples of successful international data exchange activities in each of these areas are given in Appendix C.

Laboratory Physical Sciences

The laboratory physical sciences comprise an interrelated set of disciplines that includes chemistry, materials science, physics, and the subdisciplines and applications of each of these. The primary users of most of the data generated and exchanged in these fields are other physical scientists, although data from research in chemistry, materials science, and condensed-matter and polymer physics find heavy secondary use in manufacturing and engineering applications. Recognition of potential new or changed applications often stimulates the generation of new data and concepts from basic science in these areas; the flow of stimuli as well as data runs both ways, between applied and basic sides of these sciences.

The laboratory physical sciences generate data largely from experiments, simulations, or theoretical computations.[4] (In the observational sciences, the data typically describe single, unique events, such as the weather on a particular day or the explosion of a supernova.) Although experiments in the physical sciences can be repeated, it is often the case that due to the size of the apparatus, the extent of the collaboration, the rarity or uniqueness of the test material, and the expense involved, the results of a single experiment are adopted and exchanged.[5] Instead of simply repeating experiments, scientists in these fields generally learn of a new advance and quickly use it as a steppingstone to go beyond that advance, frequently by modifying the technique or the apparatus. In the case of less complex laboratory research, scientists typically repeat the previous experiments, as much to validate the new approach as to check the previous results. The research results and the underlying data from basic experimental research are not limited by national boundaries. Information presented in an international meeting, a seminar by a foreign visitor, or an electronically circulated preprint is at least as likely as a new publication in an international journal to stimulate a new line of work. When scientists are engaged in international collaboration and exchange data that are not yet ready for publication, the national boundaries separating the collaborators are even more transparent.

Another characteristic of the physical sciences is associated with the established theoretical framework of many of the subdisciplines. The data derived from the theoretical numerical simulations in many cases look like experimental data, and often are replicated. These simulations, particularly animations, may not be part of the conventional manuscripts that report the results, but this kind of information is now exchanged globally on a variety of media. Exchanging data from simulations is a process vulnerable to the congestion problems of the Internet, described in Chapter 2, especially as the volume of such data grows.

Like all other scientific disciplines today, the physical sciences use electronic networks to coordinate, collect, compile, and distribute nonproprietary data through informal and formal means. Projects to evaluate data on particular topics, such as the thermodynamic or spectroscopic properties of a set of closely related substances, typically involve small international collaborations that communicate by Internet. More complex efforts, such as determining the "best" values of natural constants, require more formal cooperative working arrangements and regular data exchange. One such effort is the maintenance of the Evaluated Nuclear Structure Data File (ENSDF), an electronic database of evaluated data on properties of atomic nuclei and on radiation produced by decay of unstable nuclei. The database has existed in electronic form for about 25 years. An international network of individuals carries out its evaluations. Within the United States, this work is coordinated and supervised by the National Nuclear Data Center at Brookhaven National Laboratory; internationally, the International Atomic Energy Agency performs these functions. The ENSDF effort collects data from publications and other sources and then evaluates and distributes the data in a variety of formats as users call for it. Prior to the Internet, these data came to ENSDF on magnetic tapes, but now they arrive via electronic network, primarily by file transfer protocol, a convenient and widely used mode of transferring data electronically at moderately high speed. The dissemination effort is truly worldwide, with active on-line accounts in approximately 40 countries, on six continents. This system is described in more detail in Appendix C.

Physical scientists, in general, seek the most timely, lowest-cost, and most widely effective means for disseminating their results and for obtaining those of others, as long as proper citation is not compromised. Apart from proprietary data associated with commercial products, data in the physical sciences tend to be readily available, through journals, government publications, and books, and increasingly, through electronically available databases. The databases in the laboratory physical sciences may seem large in comparison with, for example, dictionaries; nonetheless, among the four areas of natural science considered in this report, the laboratory physical sciences typically have the smallest databases.

Astronomy and Space Sciences

The primary needs for and uses of data from space are in fundamental research, but there are many collateral applications, such as precise positioning, mapping of the Earth, navigation, education, and even entertainment, as the public interest in Comet Shoemaker-Levy demonstrated. Astronomy is indeed interesting to the public. As such, its data must not only be collected, but also be interpreted and made available for formal and informal educational purposes, as well as for the advancement of our knowledge about the universe.

Most data in astronomy and space sciences come from observations made from Earth's surface or from spacecraft;[6] a modest fraction of the data comes from laboratory experiments. The data from experiments, terrestrial or in spacecraft, conform closely in character to data in the laboratory physical sciences. Usually, an individual observer or observing project collects the data and distributes them to other individuals as soon as they have been taken. These data frequently have significant value to other researchers and for purposes other than those for which they were gathered. It is useful, for example, to compare data taken by different observers in different wavelength bands or to compare observations at different times in order to interpret variable objects. Hence it is important to store space science data in a form readily available to other researchers. Most astronomical data archives, which are open to all scientists, do so. Use of these archives is limited only by ease and cost of access. Consequently, this community has had to adopt efficient data management practices throughout the life cycle of the data, to permit effective access by the entire community, national and international.

Research in astronomy and space sciences is collaborative and, inherently, deeply international; it requires multinational efforts to collect data and to implement efficient transnational exchange of data. Electronic links now provide the requisite efficient communications and exchange of data. The scientific reasons for this international character include the following:

- Ground-based observatories must be located at optimal observing sites, such as mountaintops with good observing conditions, which are found only in certain countries;
- Some experiments require simultaneous observations at several points, such as long-baseline radiointerferometry;
- Only parts of the sky can be seen from any single location; and
- Some observations, such as those in the x-ray and far-ultraviolet regions, can be made only from outside the atmosphere, and hence require orbiting observatories, while others require sending probes to other planets, which creates a need for collaboration with scientists in nations that have space programs.

An economic driving force for the internationalization of space science is the high cost of large new facilities; this encourages international collaboration as a means of cost-sharing. Thus, the Hubble Space Telescope was developed and is operated as a partnership of NASA and the European Space Agency (ESA), with access available to astronomers from all over the world. The Gemini project, building two 8-meter telescopes, one in Hawaii and one in Chile, is a partnership of the United States, the United Kingdom, Canada, Chile, Argentina, and Brazil.

Even without explicit or formal collaboration, international sharing of astronomical data generally enhances the field. Recent examples include the impact of Comet Shoemaker-Levy on Jupiter, the International Halley Watch, the observations of Comet Hyakutake, and the observations of Supernova 1987a. Still less organized research projects are enabled daily by accessing archived data for historical and multiwavelength comparisons and by facilitating communication among collaborating astronomers.

Astronomers and space scientists establish research strategies and priorities for data collection in their subdisciplines. In the United States, this is usually done within the National Research Council, for example, under the decadal Astronomy Survey Committees or the Space Studies Board's planetary and space physics science strategy panels, or by NASA or National Science Foundation (NSF) science working groups or ad hoc science community studies. Other nations and international organizations develop similar research strategies, for example, the ESA's Horizon 2000, plans for the European Southern Observatory, and the international Gemini project. Such planning efforts are becoming more international and effectively identify data needs and policies in support of the projects.

Earth Sciences

In the broadest terms, Earth science data are fundamental to the discovery and creation of knowledge concerning the interactions among matter, energy, and living organisms.[7] Development of this knowledge is essential for ensuring the prospect for humanity on our finite planet in the face of rapid demographic and economic growth. Between 1820 and 1992, the world population increased 5 times and the gross domestic product per person grew 8 times, with a resulting global economy growth rate of 40 times. World trade grew more than 500 times.[8]

The best estimate at this time is that the increase in population over the next 50 years will be greater in real numbers than the increase over the last 170 years, accompanied by further large increases in economic activity and world trade.[9] This situation will bring to the fore new environmental issues and problems that will press us ever more urgently to ameliorate the impact of humankind on the environment.

Within the purview of the physical Earth sciences are natural phenomena at all spatial and temporal scales that present major scientific challenges for understanding and prediction. These phenomena include natural hazards such as hurricanes, tornadoes, floods, earthquakes, and volcanic eruptions. Besides the societal impacts associated with climate, natural hazards, and natural resources, there are numerous man-made hazards that are coupled with natural phenomena that are the subject of Earth science research. Examples include the prediction and mitigation of pollution plumes in ground water or the atmosphere (e.g., chemicals or radioactive materials), the factors involved in stratospheric ozone depletion, and the monitoring of treaties that ban underground nuclear testing (e.g., in support of the Comprehensive Test Ban Treaty).

The physical, chemical, and biological processes that shape the world in which we live are complex and interdependent. To understand them requires observations with sufficient spatial and temporal resolution and coverage to characterize the phenomena of interest and to constrain theoretical predictions that are based on conceptual or quantitative models. Therefore, the lifeblood of research in most of the Earth sciences is observational data, sometimes global in coverage, and taken repeatedly over time. Many of these data also must be integrated with data from experimental manipulations, or from other disciplines.

An example is atmospheric circulation, which controls weather over the entire Earth with significant variations on time scales ranging from hours to decades or longer, and spatial scales ranging from less than 1 km to thousands of kilometers. Weather forecasts for more than a day at a time require the rapid and repeated acquisition, processing, and interpretation of very large amounts of synoptic observations on at least a continental scale. Satellite systems that gather the necessary data have been and are being developed, but timely access to the data gathered by different organizations or countries is a major concern. Climate studies require many of the same observations as for weather prediction, but also data on the oceans, land surface, and cryosphere for the entire Earth. Therefore, international sharing of very large volumes of global atmospheric circulation data is essential for meaningful scientific investigation of past and present climates.

Scientific knowledge in the various subdisciplines of the Earth sciences has advanced to the point where important, multidisciplinary global-scale problems can be tackled with insight and scientific rigor, provided that high-quality global observations are available and that computational resources are adequate to process and interpret large and diverse data sets. Major examples of interdisciplinary and integrating research programs in the Earth sciences are the World Climate Research Program of the World Meteorological Organization (WMO) and the International Council of Scientific Unions (ICSU), the International Geosphere-Biosphere Programme organized under the auspices of ICSU, and, nationally, the U.S. Global Change Research Program.[10] These are major initiatives, begun in the 1980s to understand the driving mechanisms (both natural and human) that cause significant changes in the Earth system. These efforts involve collecting and analyzing massive data sets from Earth-observing satellites and integrating them with multiple-area or site-specific data all over the Earth, including developing countries. Significant progress in these types of complex research programs can be made only if there is effective transnational flow of data and information.

Biological Sciences

The breadth of the kinds of data in the biological sciences is probably the widest among the four areas described in this report.[11] The subjects of the data encompass types and modes of propagation of life forms, modes of provision of food and fiber, conservation of the planet's biota, public health and safety, the molecular bases of life processes, and biotechnology. Data in the biological sciences differ somewhat from those in the physical sciences, have some characteristics in common with the other observational sciences, and have some unique characteristics. Biologists have no fundamental constants or periodic table. They do share with chemists the data specifying structures of molecules, such as nuclear magnetic resonance spectra and x-ray diffraction patterns, as well as the inferred structural parameters themselves. However, many biological data specify ranges of incidence or of values of some properties. Such data require textual descriptions, which become part of the databases. Collections of such data require modes of access that are different from and frequently more complex than those that serve well in the physical sciences. Analysis by computing associations and similarities, rather than by direct, experimental, causal assessment, is characteristic of biology.

Concepts are sometimes less well defined in biology than in the physical sciences, and so clarity can be compromised when terms with even slightly different definitions, explicit or implied, are used to classify and describe what should be commonly understood data. Even the concept of "species" causes problems. For example, there are questions regarding the variabilities found within and between species and regarding whether species should be defined according to DNA sequences, with no distinctions within the species, or according to taxonomy, with differentiations made among subspecies. This issue is elaborated in greater detail below in this chapter.

Biologists use some large databases, particularly those of nucleic acid sequences that form the fast-growing genome databases. Efforts to build, maintain, and distribute the information in these databases are highly international and collaborative. Centers around the world collect data from contributing scientists and immediately share them, incorporating them as they accumulate into a coordinated database. In this respect, biologists share certain problems with the observational sciences. Proprietary concerns probably arise at least as frequently in the biological sciences as in the laboratory physical sciences or the Earth sciences, but much more frequently than in the space sciences.

A somewhat unique characteristic of many biological data, especially regarding distributions of species, is that they are very location-specific. Consequently, in order to protect fauna or flora in a given location, or the privacy of property rights of the people who live there, barriers unrelated to the research itself sometimes arise that inhibit the flow, particularly the international exchange, of biological data.

DATA TRENDS, OPPORTUNITIES, AND CHALLENGES IN THE NATURAL SCIENCES

The increasing use of electronic means for data collection, storage, manipulation, and dissemination is one of a number of broad and interrelated trends that have significant implications for access to data in the natural sciences. These trends include the following:

- Rapid growth of the body of scientific data;
- Development of large international research programs;
- Insufficient funding for data management and preservation activities worldwide;
- Decentralization of data management and distribution;
- Electronic publication; and
- Increasing use of simulations and animations as scientific data.

The discussion in this section addresses these broad trends as well as the opportunities and challenges they present. Some discipline- or field-specific issues are discussed in the next section.

Rapid Growth of the Body of Scientific Data

In every area of the sciences, both the volumes and the types of data have grown at rates unforeseeable 30 years ago. This growth has been especially rapid primarily because of vast improvements in and increasing availability of imaging detector arrays at most wavelength ranges. For example, in the Earth sciences, new technology allows data to be collected repetitively with high spatial resolution. Remote sensing systems are generating immense volumes of data that are pushing the limits of our ability to store, retrieve, and analyze those data. For instance, the introduction of ground-based Doppler radar and new satellite systems is significantly increasing the data volumes within the atmospheric sciences. Table 3.1 shows a selection of land remote sensing data sets and their anticipated volumes that are archived by the Earth Resources Observation Systems (EROS) Data Center operated by the U.S. Geological Survey in Sioux Falls, South Dakota. In seismology, new initiatives both in the United States and in other countries have resulted in continuous, broad-band digital recording at high sampling rates. Special studies using up to 1,000 sensors generate very large data sets for each experiment. Table 3.2 illustrates the actual and projected growth in data volumes at the Incorporated Research Institutions for Seismology (IRIS) Data

TABLE 3.1 Projected Volume (in Terabytes) of Satellite Remote Sensing Data Holdings at the U.S. Geological Survey's Earth Resources Observation Systems (EROS) Data Center, 1997 to 2005

Data Source	By 1997	By 1998	By 1999	By 2000	By 2001	By 2005
Landsats 1-5	120.5	122.5	122.5	122.5	122.5	122.5
AVHRR[a]	12.5	16.5	20.5	24.5	28.5	40.0
SIR-C[b]	20.0	90.0	90.0	90.0	90.0	90.0
Landsat 7	—	20.0	70.0	120.0	170.0	170.0
SRTM[c]	—	—	112.0	113.0	114.0	117.0
MODIS[d]	—	10.0	36.0	62.0	88.0	166.0
ASTER[e]	—	15.0	60.0	110.0	160.0	310.0
TOTAL	153.0	274.0	511.0	642.0	773.0	1,215.5

[a]AVHRR—Advanced Very High Resolution Radiometer
[b]SIR-C—Shuttle Imaging Radar-C
[c]SRTM—Shuttle Radar Topography Mission
[d]MODIS—Moderate-resolution Imaging Spectroradiometer
[e]ASTER—Advanced Spaceborne Thermal Emission and Reflection Radiometer

SOURCE: U.S. Geological Survey, National Mapping Division.

Management Center. Table 3.3 provides a representative sample of astrophysics data archived by NASA and demonstrates a similar trend in the space sciences.

Technology for data storage and computation continues to improve at a rate consistent with the capability to handle the rapid growth of accumulated data in the observational sciences. Scientists worldwide will have to adapt their research strategies to make effective use of these new data.[12] Although state-of-the-art projects can manage the increasingly large data volumes, perhaps with difficulty, other users, especially in developing countries, are unlikely to be able to access or effectively use such data for their own research.

Development of Large International Research Programs

As the previous discussion indicates, basic scientific research has become ever more internationalized as a result of several factors: the expanding capabilities of communication and computation networks, the capabilities for conducting high-quality science in increasing numbers of countries, and the economic driving force for sharing the high costs of large projects. These factors have led to the formation of new organizational paradigms and methods of data management. Consequences of this internationalization have been more and higher-quality science, faster progress, and ever more international involvement. These opportunities have appeared at all levels, from individual investigator and small-group science to large-scale "big science" projects.

TABLE 3.2 Summary of Actual and Projected Data Volumes Archived in the Incorporated Research Institutions for Seismology (IRIS) Data Management Center, 1994 to 2000

Data Source	Number of Instruments[a]	Data Volumes (gigabytes/year)						
		1994	1995	1996	1997	1998	1999	2000
GSN	100	1,159	2,359	3,959	6,003	8,047	10,091	12,281
FDSN	146	370	670	1,070	1,530	2,050	2,670	3,416
JSP arrays	5	1,095	2,190	3,650	5,475	7,300	9,125	10,950
OSN	30	0	0	15	58	218	498	936
PASSCAL-BB	500	1,318	2,277	3,556	5,154	7,073	9,312	11,867
PASSCAL-RR	500	542	885	1,341	1,912	2,597	3,397	4,310
Regional-Trig	500	150	290	490	730	1,030	1,390	1,755
TOTAL	1,781	4,634	8,671	14,081	20,862	28,315	36,483	45,515

NOTE: Abbreviations are as follows:
GSN—Global Seismic Network (IRIS)
FDSN—Federation of Digital Seismic Networks
JSP—Joint Seismic Program (with the former Soviet Union) (IRIS)
OSN—Ocean Seismic Network
PASSCAL-BB—Program for Array Studies of the Continental Lithosphere—Broadband (IRIS)
PASSCAL-RR—Program for Array Studies of the Continental Lithosphere—Regional Recordings (IRIS)
Regional-Trig—Regional Triggered Recordings

[a]Projected for the year 2000.

SOURCE: IRIS Data Management Center, private communication, 1994.

Recent years have seen several new multibillion-dollar international projects, and multimillion-dollar international efforts have become almost commonplace. These "megaprojects" or "megascience" programs have a number of common characteristics. They require long-term funding commitments; they may necessitate the building of new large facilities or instruments, which then require large expenditures for operating funds; they typically involve teams of researchers working on different aspects of the project, with the consequent requirement for international communication and data exchange; and, with the current state of technology, their scientific objectives cannot be fulfilled by using a smaller-scale research format.[13]

In 1991 the U.S. Congressional Budget Office identified 80 projects funded by the U.S. government that each cost at least $25 million (in 1984 dollars) during the period from 1980 to 1986.[14] Many of these involved significant international participation. In contrast, there were only a handful of such nonmilitary large-scale research projects in the 1950s and 1960s.

TABLE 3.3 A Representative Sample of NASA Astrophysics Archives, by Satellite Mission

	High Energy Astrophysical Observatory 2	International Ultraviolet Explorer	Infrared Astronomical Satellite	Hubble Space Telescope	Compton Gamma Ray Observatory
Data type	X-ray data	Ultraviolet data	Infrared data	Optical/Ultraviolet data	Gamma-ray data
Year of launch	1978	1978	1983	1990	1990
Duration	2.5 years	Ongoing	300 days	Ongoing	Ongoing
Total data volume (gigabytes)	~100	~100	~150	~5,500 by year 2005	~10,00 by year 2000
Data center	Einstein Observatory Data Center, Cambridge, Massachusetts	National Space Science Data Center, Greenbelt, Maryland	Infrared Data Analysis Center, Pasadena, California	Space Telescope Science Institute, Baltimore, Maryland	National Space Science Data Center, Greenbelt, Maryland

SOURCE: NASA; reprinted from National Research Council (1995) *Preserving Scientific Data on Our Physical Universe: A New Strategy for Archiving the Nation's Scientific Information Resources*, National Academy Press, Washington, D.C.

For the purpose of this discussion, it is useful to identify several types of large international scientific projects and programs:[15]

- *Experimental facilities* for neutron beam, synchrotron radiation sources, lasers, high-energy particle physics, high-field magnet laboratories, and fusion experiments.
- *Fixed observational facilities* such as optical and radio ground-based telescopes, environmental remote sensors (e.g., lidars and radars), and deep ocean drilling projects.
- *Space science observational satellites*, including space telescopes for astronomy and astrophysics, space physics observatories, and planetary missions.
- *Earth observation satellites* for collecting data about Earth's atmosphere, oceans, land surface, and geophysics.
- *Distributed observational programs* that collect data in many different locations as part of an internationally organized research program. Examples of this include the global seismic network, the International Geosphere-Biosphere Programme, the Human Genome Project, and the new Biodiversitas project.

It is the latter two types of scientific research initiatives that pose the greatest data management challenges for effective international exchange, as discussed below in this chapter.

Insufficient Funding for Data Management and Preservation

Despite the vast increases in recent years in the amount of data collected and stored, and the very large augmentations in the funding allocated to new observatories and experimental facilities in ambitious international research programs, there has not been a commensurate increase in the funding for data management and preservation. At the same time, the costs associated with data retention and distribution are typically far less than the costs of reacquisition (in those cases in which reacquisition is even physically possible). Although the committee did not perform comprehensive research on the actual funding levels worldwide, the members of the committee believe this to be a problem common to most disciplines, in many research programs, on both domestic and international levels.

In the laboratory sciences, funding agencies focus their support on research and tend to overlook the fact that data compilations and data access are key to progress across the research spectrum. Furthermore, the sheer volume of data now available makes it increasingly difficult for individual compilers in the tradition of Ptolemy and Beilstein to fill this need as a pro bono activity, or for the work to be done as merely ancillary to some funded research program. In addition, society fellowship and award committees generally do not place much value on the contributions their applicants may make to the infrastructure of science in the form of data compilation, organization, and evaluation work. Funding agencies have an opportunity to enhance the international aspect of these activities by supporting scientists from developing countries who tend to be well-educated but underemployed. Such a policy could be very cost-effective in that it would not require high initial capital costs for facilities, but only labor and data access costs.

The funding situation in the observational sciences tends to be even more difficult, with potentially more significant negative effects, given the data-intensive nature of such research. Experience indicates that scientists associated with new observatories get much more support than those handling data from old ones, even though the payoff from optimal utilization of existing data sometimes is greater. For instance, according to figures supplied by NOAA, the agency's budget for its national data centers in FY 1980 was $24.6 million, and their total data volume was approximately 1 terabyte. In FY 1994, the budget was only $22.0 million (not adjusted for inflation), while the volume of their combined data holdings was about 230 terabytes! During this same period, the overall NOAA budget increased from $827.5 million to $1.86 billion, mostly to fund the acquisition of new observational data.

An example of insufficient funding for data management and preservation is the National Land Remote Sensing Satellite Data Archive at the EROS Data Center. This national archive was established by Congress[16] in 1992 and endorsed by the 1996 National Space Policy without the provision of any new funding to support its expanded mission.

Although improvements and significant cost reductions in data storage and processing technologies have enabled government data managers to keep up with the demands in most cases, the pressures from chronic underfunding occasionally have led to ill-considered attempts to commercialize or privatize the data management and dissemination functions (as discussed in Chapters 4 and 5). In other cases, these financial difficulties have led to inadequate preservation and access provisions, which sometimes result in the partial or total loss of irreplaceable data sets.[17] The challenge is to develop data management and archiving infrastructure and procedures that can handle the rapid increases in the volumes of scientific data, and at the same time maintain older archived data in an easily accessible, usable form. An important part of this challenge is to persuade policymakers that scientific data are indeed a precious resource that should be preserved and used broadly to advance science and to benefit society.[18]

Decentralization of Data Management and Distribution

Data collection, management, and distribution over the long term depend on a variety of institutions, that is, organizations that transcend the interest of individuals or ad hoc groups of scientists. These institutions have various roles, missions, and funding responsibilities that affect use of and access to scientific data. Many have direct responsibilities or objectives regarding data generated by publicly funded research, including, in particular, the following institutions:

SCIENTIFIC ISSUES   63

- Government science agencies and academies of science;
- Intergovernmental scientific organizations and coordinating bodies;
- Publicly funded data management institutions, including data centers and libraries;
- Publicly funded research institutions, primarily in academia;
- National and international nongovernmental organizations, such as scientific and engineering societies, library associations, and information industry associations;
- Commercial publishers; and
- Governmental policymaking and regulatory bodies.

The information technology revolution has changed the roles of some of these institutions and brought about the establishment of new entities. For scientific data management at both the national and the international level, the technological changes have supported the development of organizations with the following attributes:[19]

- *Widely distributed responsibility.* New telecommunications, data management, and standardized technology is leading to highly reliable distributed data management capabilities. The growing availability of information technology professionals (along with the lower technical skill levels actually needed by end users) is enhancing the ability to distribute data more broadly and increase user participation. Such distribution of data and their ownership (whether actual or implied) by user groups improve the utility of the data and help create important support for long-term retention.
- *High-value peer-to-peer communication.* With on-line access to data and people, a variety of new collaborative relationships can develop. Information can be broadcast to interested individuals in a timely fashion. Data can be provided directly to field researchers to focus new data collection. Physical proximity and formal lines of communication are no longer vital to effective organizational operation. Indeed, closed, highly structured organizations often will be uncompetitive or unable to take full advantage of innovation.
- *Specialized data functions.* When resources and capabilities are distributed, some specific locations can make an effective contribution by specializing. Specialized groups can be created in a scientific discipline or in some aspect of data management, archiving, or standard setting. Such centers can achieve significant economies of scale, reducing overall costs while enhancing the effectiveness of certain functions for the benefit of all.

The National Nuclear Data Center (NNDC) is one of many examples of the evolution from sole source to distributed network. Under an international agreement, NNDC at Brookhaven National Laboratory is the U.S. source for the international distribution of evaluated nuclear data files (see Appendix C). When the primary forms of distribution were hard-copy books or published tables, the sole source was obvious—the printed pages generated at NNDC. When on-line access to the databases became available in the mid-1980s, the files had to be mirrored for easier access overseas. With the advent of the World Wide Web, individuals gained the ability to manipulate the databases, which today still reside at NNDC, using overlay programs that (physically) reside at another data center, possibly in Europe.

In the future the user probably will be unaware of where the data file being accessed physically resides and will be able to link to the journal article (published by a commercial publisher, for example) in which the data originally appeared. The electronic journal article could have links to the original data tables of the authors. NNDC has evolved from a collector of evaluated data files, which were formatted into camera-ready pages for hard-copy publication, to a center that maintains on-line access to a few databases via a variety of overlay programs (no longer is there a single, static format). It is one of many centers around the world, now common in most scientific disciplines, that develops new, electronic forms of networked data dissemination.

Electronic Publication

The development and acceptance of electronic networks as a means of communicating, searching for data and information, and accessing information rapidly and directly have driven the increase in electronic publications of all types, and scientific publications in particular.[20] Although not all publishers of scientific journals are moving to completely electronic form, there is a distinct trend to provide alternative paper and electronic versions of many publications. For example, the American Institute of Physics is working to provide its library clients by early 1997 with electronic access to every one of its journals to which the library subscribes. In addition, it now offers some of its journals in CD-ROM form as a space-saving alternative or supplement to its subscribers.[21] The Institute of Physics in the United Kingdom also provides subscribers with electronic access to all 33 of its journals.[22]

NASA is sponsoring an all-electronic peer-reviewed journal, *Earth Interactions*.[23] It is only available electronically and allows color representations of phenomena, including time-lapsed video clips of observations to show time variations. Mathematical calculations on subsets of original data can be carried out by the reader as well, providing both in-depth understanding of the material and a check on the validity of the author's results. This new journal is the product of three professional societies—the American Meteorological Society, the American Geophysical Union, and the Association of American Geographers—with additional support from the Ecological Society of America and the Oceanographic Society of America. These societies have a combined membership of approximately 45,000, and so this form of electronic publication will soon be available to a substantial segment of the Earth sciences community. Submission, editing, and peer review will all be done electronically.

There is increased attention to electronic publication of astronomical research papers and data as well (see Box 3.1). Most conference proceedings are collated from electronic submissions in standard (e.g., TeX, LaTeX) formats. Abstracts of papers in most space science disciplines are now available on-line,[24] and several journals are publishing electronic versions.

This trend is likely to accelerate and to open new opportunities for communicating research results to all scientists. Electronic publication will not just replace paper, however—it will alter the sociology of science. Writing, refereeing, and reviewing of a publication are now discrete and strictly ordered events, but they need not be in the electronic world. There, annotation, critique, elaboration, and revision can all go on iteratively and indefinitely, and in some instances no doubt they will. Some publications likely will become "living documents," under revision until they are no longer of interest. Even though our current social norms for attribution are based on the static publication model, it is doubtful that the scientific community would retain that model in order to preserve these norms. The value of a dynamic discourse is too great.[25]

Many electronic journals will not be "printable" in any meaningful sense. It is not just that they will contain motion and sound, but will incorporate also rich contextual links to the primary materials. Clicking on a graph will give the reader access to the data on which it is based, allowing alternative models and interpretations to be explored. A related important benefit of electronic publications is that results based on observations and modeling can be checked and validated by both reviewers and readers; restrictions on article length in paper journals and limited access to original data and software currently preclude any meaningful checks of the validity of published results based on observational data. A "copy" of the bits in an astronomical "plate" is as good as the original.

Publications also will become "active" agents, rather than passive stacks of paper. The term "program" has some of the wrong connotations, but nonetheless future publications will include executing programs—not ones that can be executed, but that are executing—autonomously gathering data, making predictions, becoming richer and more valuable as time passes.

In short, the Internet and World Wide Web are far more rapid and enabling means of communicating results, ideas, and other aspects of research than paper publications. Many changes in the conduct and dissemination of scientific research, from the individual to the international scale, may be expected to arise from these developments.[26]

Increasing Use of Simulations and Animations as Scientific Data

Related to the trend in electronic publishing is the increasing use of simulations and animations in research.[27] Large-scale computation arose in part from a

> **BOX 3.1**
> **Space Science Data and Electronic Publishing**
>
> The space science community has been at the forefront of electronic publishing. Scientific societies, such as the American Astronomical Society (AAS), have been leaders in this type of information exchange. For example, AAS collects all of its meeting abstracts electronically and publishes the AAS Job Register on-line. In addition, AAS pioneered the development of effective electronic journals with its on-line publication of the Astrophysical Journal. What started out as an experiment has turned into a success story in electronic publication.[1]
>
> Supported in part by a grant from the National Science Foundation, the AAS in cooperation with the University of Chicago Press first developed an electronic version of the Letters portion of the *Astrophysical Journal* (eApJ).[2] Produced in two versions (HTML for screen reading and PDF for local printing), this journal goes well beyond the electronic delivery of paper manuscripts typical of most "electronic" journals.
>
> The eApJ has references tied into the NASA-supported bibliographic database maintained by the Astrophysics Data System (ADS; see <http://adswww.harvard.edu>), which provides abstracts of most references and is developing an archive of page images of several of the most useful astronomical scholarly journals.
>
> The eApJ uses URNs instead of URLs (names instead of locations) as link targets, and so the links will remain valid indefinitely. Both the ADS and the eApJ use a standardized notation for naming articles, which enables links and pointers to be generated automatically during the publishing process. As part of the sophisticated set of links associated with this journal, the eApJ includes a capacity for forward referencing whereby each article carries with it an updated set of references to articles that refer to it—an automated electronic citation service. Before the full *Astrophysical Journal* came on-line in November 1996, the AAS made arrangements to establish mirror sites in Great Britain, Europe, Australia, and (possibly) Japan to ensure relatively rapid response times.
>
> The success of the eApJ is propelling the astronomical publishers to bring their literature on-line rapidly. Over 95 percent of the world's peer-reviewed astronomical literature is expected to be on-line by mid-1997. Standard protocols, conventions, and procedures will be absolutely critical if this networked system of literature and data is to be effective for the working scientist.
>
> Electronic publishing in this area has also enhanced data access and archiving. The Astronomical Data Center, located in Strasbourg, France, has an agreement with the publishers of *Astronomy and Astrophysics* "to provide all data files from their publications." The Astronomical Data Center at Goddard has a similar arrangement with the AAS, which includes *Icarus* and the publications of the Astronomical Society of the Pacific, as well as AAS publications. This type of arrangement "permits the two centers to archive a major portion of the international astronomical data without individual requests to the authors" of journal publications.[3]
>
> ---
>
> [1]Response by Peter Boyce, American Astronomical Society, to the committee's "Inquiry to Interested Parties" (see Appendix D).
> [2]The full journal is now available on-line at <http://www.journals.uchicago.edu/ApJ/>.
> [3]Response by Nancy G. Roman, NASA Goddard Space Flight Center, to the committee's "Inquiry to Interested Parties" (see Appendix D).

need to simulate reactive hydrodynamic flows. Such problems still help drive the development of increasingly powerful computers. However, during the past three decades, simulations have become integrated elements of the toolboxes of experimentalists and theorists in many of the physical and biological sciences.

The increasing importance of modeling and simulation is evidenced for the materials science field by the recent (1992) inauguration of two new journals devoted exclusively to this topic: *Computational Materials Science* (Elsevier) and *Modeling and Simulation in Materials Science and Engineering* (Institute of Physics Publishing). Materials data modeling encompasses two quite different areas: materials R&D (both theoretical and experimental) and data handling and application activities (continuum level design calculations, process modeling, service behavior modeling, and compression, extrapolation, and interpolation of data). Other research areas in which computer simulations have become standard are in the design of optics for electromagnetic radiation and of beams of electrons and ions; flow of fluids; folding of protein molecules; interaction of enzyme molecules with their substrates, the species on which they act; melting and freezing, at the atomic level; the motions of individual atoms during reactive collisions of molecules; and collisions of gaseous atoms and molecules with surfaces.

Many simulations require repeated solution of equations of motion of the system by computer. These equations may be simple or complex, but however simple they are, the ability to solve them over and over, many millions of times, as the system they describe evolves, is a consequence of the power of electronic computers. The simulation results may be reduced to only a few summary numbers, which was the usual practice in the early years of computers. Now it is common for the results to include numerical information about entire time histories, information that can be put into tables and graphs.

Perhaps the most dramatic advance in simulations, however, has been the use of graphics, particularly animations. The information in an animation can give insights into a scientific phenomenon that could not be guessed from individual snapshot images or numerical indicators. Time may serve as a surrogate for a spatial dimension, allowing the investigator to visualize the behavior of a function of three independent variables. Animations are useful when an investigator uses a preconception to decide what indicators would be best to compute and it turns out that the situation does not correspond to that preconception. For example, one study examined the high degree of solid-like, cooperative, and collective motion of most, but not all, of the atoms in the supposedly liquid surface layer of a cluster of atoms. Instead of an amorphous swarm of atoms swirling on the surface, the outer, "molten" layer of the cluster showed organized, collective (but loose) vibrations by all but a few of the surface atoms. All the quantitative indicators showing liquid-like character arose from a few atoms displaced from the surface so that they were free to float just outside it, in almost-free fashion. The consequence of seeing the animation was the construction of a theoretical model very different from the one the investigators had been planning to use. In short, animations have become real tools of research, not just pedagogic devices.

The data in animations cannot yet be stored or exchanged in conventional journals. However, electronic storage and transmission over the Internet and the World Wide Web make it possible for scientists to share not only their tables and still figures, but also their animations. A few published papers include references to Web addresses that provide animations of material discussed in the papers. Such data are still in a form that is primitive compared with what one can foresee. Now it is possible to play an animation, even a multiwindow display that shows several characteristics evolving simultaneously, and sometimes to stop the motion to study an individual frame. In the future, it will be possible for the viewer to stop an animation and examine the image or images from all sides, perhaps even to carry out manipulative operations on that image that correspond to simulations of physical processes. The capacity for such data manipulations will advance as the bandwidth available for data transmission increases, as data storage becomes cheaper and faster, and as the software for generating, storing, and displaying animations and more elaborate time histories becomes more user-friendly. It is already inevitable that scientists will generate image sets for molecular phenomena analogous to the computer-based "tours" of towns and cities that allow the viewer to choose any path through the area. For example, it is possible for a pharmacological researcher to "fly" a molecule of a potential new drug to a conjectured target receptor site, say in the brain, to judge whether it is a topological fit.

The capability to share images from animations has a number of advantages. However, images require more storage space, and the user must have considerably larger bandwidth capabilities to allow electronic transmission of the animations than typically are required for exchange of numerical or symbolic data.

DISCIPLINE-SPECIFIC DATA ISSUES

Over the past two decades, the National Research Council and other groups have issued numerous reports that have addressed scientific management issues for digital observational data in the Earth and space sciences.[28] More recently, several studies have examined such issues in the biological sciences.[29] Most of these reports have focused quite narrowly on the data management problems of specific disciplines or agencies; however, many of their recommendations have broader validity and may be applied to other disciplines and institutions in the observational sciences in the international context.

As noted above in this chapter, the very large scale environmental observational research programs in the Earth and biological sciences pose the greatest data management challenges and the most difficult public policy issues. This is so not only because of the complexity of the scientific questions in those disciplines and their data-intensive nature, but also because of their inseparable linkages to major socioeconomic issues, the potential for private-sector exploitation of the data and related research results, and their relevance to other major government concerns such as national security, trade, and foreign policy objectives. Although the focus of this study is on issues in the transnational flow of scientific data for use in basic research and not on the role of scientific data in these other, much broader contexts, it is at the interface of scientific use of data and their broader potential applications that the most vexing public policy issues arise—a topic addressed in some of the discussion below in this report. The sections immediately following, however, focus on some of the more specific discipline-related challenges and opportunities in providing broad international access to scientific data.

Observational Environmental Sciences

Measuring and Monitoring Systems

The data now available on a global basis are inadequate to document and understand many environmental and health problems, or to anticipate problems that may arise because of the increasing influence of human activity. What is required is a comprehensive and long-term effort to observe, understand, assess, and predict the global environment—a World Environmental Watch.

Significant progress has been made toward this end, beginning with a series of National Research Council studies in the 1980s that outlined the rationale and data requirements for a new branch of scientific inquiry called Earth system science.[30] This in turn led to the formation of ambitious international global research programs, such as the International Geosphere-Biosphere Programme and the U.S. Global Change Research Program. A set of international complementary observing systems has been proposed, elements of which are in various stages of deployment and development. Included in this set are the Global Climate Observing System (which includes the World Weather Watch and the Global Atmospheric Watch), the Global Ocean Observing System, and the Global Terrestrial Observing System.[31] These internationally coordinated efforts will integrate observations from multiple satellites and airborne and in situ sensors deployed worldwide.

The Global Ocean Observing System and the Global Terrestrial Observing System are in earlier stages of development than the World Weather Watch and Global Atmospheric Watch and the associated Global Climate Observing System (see Appendix C for a description of the World Weather Watch). However, for Earth's land surface, substantial observing systems exist within many countries, and for the oceans, observing systems have been developed by countries to a considerable extent for the coastal oceans and to a limited degree for the open ocean through international collaboration.

One of the oldest international observing systems (over 100 years old) is the volunteer ship observing program coordinated by the WMO and the Intergovernmental Oceanographic Commission. Through this program involving thousands of ships from many countries, weather and sea surface temperature observations have been available to all countries in real time. It also has provided data on the climatology of the ocean area for many decades and still does in conjunction with meteorological satellites.

More recently, an extensive observational network for measuring the upper ocean was put in place in the western Pacific Ocean as part of the Tropical Ocean Global Atmosphere Program. The ocean data from this network and the data from the World Weather Watch Global Observing System provide the basis for the development of atmosphere-ocean coupled models, which have formed the foundation for experimental forecasts in seasonal to interannual predictions. Extensive plans also have been formulated for the components of the Global Ocean Observing System to support the study of climate more generally.

The Global Terrestrial Observing System has several major components for the land surface, surface and ground water, and seismology. Most nations have developed hydrologic observational networks for both the surface and the subsurface water. River stage observations are taken in most countries, both for flood forecasting and for water resource management. The further development of these observational systems is essential if the nations of the world are to cope with the wide range of environmental changes that are occurring and can be envisaged.

In the biological environmental sciences, monitoring systems are much less fully developed than in the Earth sciences. Carefully planned and coordinated global monitoring systems for new and emerging diseases and ecological monitoring and biodiversity surveys are needed. An epidemiological system now in place determines which strains of influenza virus are emerging each year. The composition of each year's vaccine depends on effective monitoring and early warning. Recent outbreaks of Ebola virus in Africa indicate the need for more monitoring information that combines epidemiological and ecological data.

An example of a lack of ecological monitoring comes from consideration of the world's island ecosystems. We know a great deal about animals and plants in special habitats such as the Galapagos Islands, but essentially nothing about their microbiota. We do not know the similarities and differences in microbial ecology between the Galapagos Islands and, for example, the Cape Verde Islands, despite their geological similarities.

Given the nature of the regional and global problems and the interdisciplinary nature of the environmental and health sciences, research on a specific problem often requires the use of data from several observing systems. Therefore, important requirements are observational consistency in space and time, with accurate georeferencing to the maximum degree possible; thorough documentation of data attributes; and substantial institutional commitments to the long-term continuity of key observational programs.

Quality Control and Assurance

Quality control operates smoothly and almost transparently in those sciences in which experiments are readily reproducible or lead to subsequent experiments that validate the original ones. In the observational sciences, implementing effective quality control for data requires the use of an audit trail system that includes anomaly detection, reporting, and correction, as well as the rigorous refereeing of manuscripts for publication.

Quality assurance, the mechanism used by management to assure that the quality of work is as claimed by those doing it, typically plays a far smaller role in basic science than in applied science and especially in manufacturing. However, one can interpret any mechanism to assure scientific integrity as a kind of quality assurance procedure. This concept would thus include the mechanisms to detect and investigate scientific fraud. Such "quality assurance" efforts are carried out in universities and at the National Institutes of Health, for example.

In recent years, some organizations, such as the Carbon Dioxide Information Analysis Center (CDIAC) at Oak Ridge National Laboratory have devoted significant efforts to producing high quality global Earth science data sets whose accuracy and reliability have been determined, accompanied by the descriptive (metadata) documentation needed for their use. The CDIAC has quality-assured and documented several key global change databases on such diverse topics as concentrations of carbon dioxide and other greenhouse gases in the atmosphere, carbon fluxes from the terrestrial biosphere to the atmosphere resulting from changes in land use, carbon chemistry in the oceans, and long-term climate trends in the United States.[32] These value-added data sets are certified as valid by the primary users who collected the data, or by those who subsequently carried out the quality-control checks of the data. This is a somewhat costly, but successful, approach for assuring secondary users of the quality of relevant data sets.[33]

Preservation of Historical Data Sets

The trend toward bigger, more complex, and more expensive facilities and programs in the observational sciences, and toward attendant international collaboration, has brought about greater attention to and incentives for effectively archiving data. It also has encouraged the development and maintenance of a curatorial infrastructure necessary to manage the data better, to provide more uniform processing and documentation, and to make retrospective data more easily accessible and usable.[34]

Research using archived data has grown in scope and importance, especially in enabling the comparison of observations taken at different wavelengths and at different sites. In the space sciences, there are efforts to coordinate data catalogs and indices, facilitating discovery of what data are available (e.g., NASA's Astrophysics Data System and its extragalactic database, SIMBAD). Space astronomy in the United States has been at the forefront of archiving and distributing data electronically. Archives are an integral part of all U.S. space astronomy missions and are now typically being planned for ground-based observatories and for other countries' space missions as well. The archiving technologies are openly available and shared, and data sets from new and different observations are incorporated increasingly in existing archives. Comprehensive catalogs and good user-access tools are recognized as very important, as are properly maintained and preserved duplicate data sets. Some major archives are also duplicated at different sites to reduce communication loads and to promote innovation and allow different uses.

In the Earth sciences, the study of Earth processes involves time-dependent behavior over time scales ranging from seconds to millions of years. For relatively short time scales (years or less), observational data from a common observing platform may be available from a single database. For longer time scales (decades to centuries to many thousands of years), it is necessary to scrutinize all of the retrospective observations available and to use proxy data preserved in the geologic record or in written records. Research on global change and on natural hazards, for example, whose goal is improved prediction of future conditions or events, depends heavily on accurately reconstructing the record of the past. Box 3.2 provides several examples of interesting data reconstruction projects in China.

Because there has been an increasing awareness of the great value of retrospective Earth science data, some conscientious efforts to rescue and preserve older data are being made both nationally and internationally. In the United States for example, the National Climatic Data Center and the National Geophysical Data Center have devoted time and resources to data rescue in recent years and now have a policy of transferring all their digital data holdings to new storage media at least once every 10 years.

Many specialized observational databases requiring long-term retention exist in biology as well. These include such diverse subjects as agricultural records of many types, including experimental field tests going back to the last century; museum, zoo, herbaria, and microbial culture collection records; hospital and other medical records; ecological data; breeding histories of domestic animals and plants; macromolecular sequences and their accompanying annotations; taxonomic treatments, toxicological information; folk medicine; and characterization of biological products such as food, fiber, and fine and bulk chemicals. Some of these data are in computers, but require normalization for consistency and readability. Others could be transformed into machine-readable form to make them generally accessible. These two tasks are labor intensive and require a large component of highly skilled labor. Such undertakings require careful prior evaluation for potential worth. Selective evaluation and support is necessary not only to enhance the intellectual effort to maintain existing databases, but also enable the creation of new ones.

Once the primary data are analyzed and used to publish research results, the

> **BOX 3.2**
> **Examples of Digitizing Historical Environmental Data**
>
> In the People's Republic of China the following efforts have led to the digitizing of these historical records, among others:
>
> - A joint study between the Chinese Academy of Sciences and the U.S. Department of Energy used proxy records for grain harvests to extend the annual rainfall data for Beijing back to 1260 A.D.[1]
> - Computer analysis of ancient Chinese sunrise eclipse records showed that the length of the day was 70 milliseconds shorter in 1876 B.C. than now.[2]
> - Information on medicinal plants from Chinese pharmacopia from thousands of years ago, as well as other folk medicine records, are being computerized in an effort to identify active ingredients that might form the basis for modern drugs.[3]
>
> ---
>
> [1] F.A. Koomanoff, Ye Duzheng, Zhao Jianping, M.R. Riches, W-C. Wang, and Tao Shivan (1988), "The United States Department of Energy and the People's Republic of China Academy of Sciences Joint Research on the Greenhouse Effect," *Bull. Am. Meteorol. Soc.*, 69:1301.
> [2] K.D. Pang, K. Yau, H.H. Chou, and R. Wolff (1988), "Computer Analysis of Some Ancient Chinese Sunrise Eclipse Records to Determine the Earth's Rotation Rate," *Vistas in Astronomy*, 31:833-847.
> [3] Senliang Li (1990), "Data Acquisition of the Chinese Medicinal Plant Database," presentation at the CODATA International Conference, Columbus, Ohio.

authoring scientist may be reluctant or inattentive about placing the unpublished primary data in a publicly accessible database or archive. Rather, the scientist is likely to concentrate on creating new data to be interpreted and summarized for additional printed publications. No incentives exist in most biological disciplines to encourage the contribution of primary data to databases. The few exceptions mostly involve data on biological macromolecules such as proteins, DNA, RNA, and complex carbohydrates. No crediting mechanism, however, adds to professional standing in the same way as a printed publication. The result is the loss of a great deal of useful data.

The long-term retention of biological databases also is being funded and managed in a haphazard, uncoordinated fashion throughout the world. This chaotic situation is unnecessary. The worldwide cooperation in establishing DNA and RNA genetic sequence databanks pointed out in Appendix C demonstrates what can be done. The world's information science community, together with the world's biologists, now have the combined skills and much of the infrastructure to preserve and to make basic biological information resources broadly available.[35] The scientific base and technology exist to produce much needed information structures that are the biological equivalent of the global weather monitoring system. The critical biological problems the world's population faces in such areas as medicine, agriculture, sustainable ecology, food production, and water quality know no political boundaries and require effective information transfer and access to archival or baseline data.

The fact that observational data are unique and not reproducible leads to the conclusion that they should be preserved as part of the historical record of the dynamic behavior of Earth and its inhabitants. The intrinsic, long-term value of observational scientific data was emphasized and discussed in detail in a recent National Research Council (NRC) study.[36] A major point made there was that such data are an invaluable national (and by inference international) resource that should be preserved and utilized to advance the state of knowledge about our natural environment.

If this view becomes widely accepted by national and international organizations and governmental bodies, the traditional practices of allowing older data to become increasingly inaccessible or destroying them will be supplanted by policies and procedures to preserve retrospective data in accessible and usable forms. It should be noted that the data volumes of all previously collected data in a given area of the observational sciences typically are modest or insignificant in comparison with the volumes that the current data collection systems produce; if there were a policy of preserving older data indefinitely into the future, all prior data would be transferred to new storage media in compatible formats as new storage and retrieval technology is adopted.

Despite the ability of data storage and computational technology to keep pace with the data volumes being generated, there may be instances in which sufficient resources may not be available to preserve all the useful data from a research program or a science agency. In the unlikely event that a suitable long-term repository cannot be found, the decision regarding what data to purge should be made by representatives of the primary discipline and other major user groups.

Data Integration for Interdisciplinary Research

The improvements in technological capabilities have led to new opportunities to address important scientific problems that earlier were either obscure or considered intractable. Although a considerable amount of research in the observational sciences continues to be done by individual investigators in specialty areas, there are now many more multidisciplinary, multiinvestigator studies involving complex natural processes in space and time. Attendant to this evolution is the need to access diverse types of complementary observations made by many different scientists and organizations around the world.

The large international initiatives in global change research mentioned earlier provide good examples of programs in which success is critically dependent on the transnational flow of scientific data and information. The observational data necessary to obtain meaningful results in most areas of investigation include in situ (point or local) measurements, regional observations using various observing platforms (e.g., balloons, remotely piloted vehicles, airplanes), and satellite remote sensing.

A major Earth science initiative involving significant transnational flow of scientific data is the International Decade of Natural Disaster Reduction.[37] The focus of this initiative is on understanding the dynamic processes that cause major natural disasters (e.g., earthquakes, volcanic eruptions, tsunamis, floods, hurricanes, tornadoes) and mitigating their effects through enhanced prediction capabilities and precautionary safety measures. All of these phenomena occur globally, and their study involves comparisons of many different types of observations. Commonly, cooperative studies are carried out by scientists from several countries and through agreements among countries or institutions. International sharing of a wide variety of observational data, including demographic data documenting human and economic effects, is essential for improving our knowledge of these natural hazards.

A particularly challenging problem is accessing and merging relatively sparse, lower-resolution, retrospective observations with the higher-resolution current observational data to document changes occurring in the environment. A recent NRC report, *Finding the Forest in the Trees: The Challenge of Combining Diverse Environmental Data*, provided possible solutions for integrating multiple environmental data sets at different spatial and temporal scales. This report considered in detail six case studies to elucidate the problems of interfacing diverse types of geophysical and ecological data to address important environmental problems in a global context. The lessons learned from these case studies provided the basis for a set of recommendations to overcome barriers deriving from the data themselves, from users' needs, from organizational interactions, and from system considerations. The committee endorses those recommendations and incorporates them here by reference.[38]

The interfacing of several different environmental data sets in a single research project can be difficult because the data layers are not coregistered to a universal template and therefore do not "stack" perfectly. The misalignment of only a few tenths of a degree in longitude or latitude creates major problems and leads to misinterpretations. For example, regions of the world with complex coastlines are very difficult to study when co-registration of data layers is not perfect.

An important tool is Geographic Information System (GIS) software, which enables all types of data to be correlated or compared geographically. This capability greatly facilitates multidisciplinary research that involves many different types of observational data and disparate scales of sampling ranging from point measurements to repeated, synoptic, high-resolution satellite imagery. Until recently, most GIS software was geared toward spatial data, but now the time dimension is being incorporated more formally through four-dimensional assimilation techniques. GIS is being used routinely both for fundamental scientific research and for applications in the Earth sciences, although many problems

> **BOX 3.3**
> **Barriers Encountered in International Environmental Assessments Using Geographic Information Systems**
>
> "I have been working for five years on an EPA-supported project developing geographic data and geographic information systems for environmental assessment on the Mexico-U.S. border. This effort has required the acquisition, verification and enhancement of various types of geographic, earth science and demographic data from Mexico, the US and several international agencies.
>
> "During this binational project, I have encountered several major issues associated with geographic data. At the conceptual level, these issues involve the cognitive representation of the landscape used to capture the data and to represent it in the digital domain. Additionally, standardization of data acquisition methods, geographic scale, resolution, spatial accuracy, feature (attribute) definition and metadata presentation are some of the technical issues that lead to a lack of comparability and impair maximum utility in a binational setting. I have found that the definition and resolution of the conceptual and the technical issues are influenced greatly by the differences in culture (scientific, political, philosophical), language, economic development, etc. Seldom is the reality of the earth's surface represented identically in the mental maps of two individuals. The translation of this perceived reality to digital maps distorts the variation even more. Unlike other types of data, the wide variety of distinct applications for individual geographic and earth science data sets compounds the potential for misunderstanding and misuse by various user groups."
>
> SOURCE: George F. Hepner, University of Utah, personal communication, January 18, 1996.

remain.[39] Box 3.3 summarizes some of the barriers encountered in an international environmental assessment project using GIS.

Documentation to Support Secondary Users of Observational Data

As discussed above, research in the observational sciences increasingly involves the integration of multiple, diverse data sets, most or all of which were not collected by the end users. The primary researchers who collect the data often do not make the effort to include the documentation that secondary users need. These secondary users,[40] who frequently are less knowledgeable or technically sophisticated, must have sufficient information about the data (i.e., metadata) in order to avoid possible misuse and misinterpretation. Of course, if the data are used improperly, incorrect results and interpretations are likely to result, and these, in turn, may be propagated through the scientific community of secondary users, thereby spawning still more erroneous interpretations.

Therefore, a key component of effective international and interdisciplinary use of scientific data is the associated metadata that describe where and how the data were collected, the calibrations that convert raw data into physical units, corrections that have been made, the quality and reliability of the data, the data format(s) and any other information or caveats concerning the proper use of the data.[41]

For example, research on global change is in large part being carried out by secondary users. Past observations of the temperature at Earth's surface, gathered at many locations globally for weather, agriculture, or environmental studies, provide a long-term record that is being investigated to determine whether global warming is taking place. In other cases, one type of observation may be the proxy for another parameter that has not been adequately observed (e.g., cloud cover as a proxy for precipitation). A remote sensing example is the original Landsat, which was designed for looking at vegetation, crops, and other agricultural purposes but has been extensively used for geological studies; likewise, the Seasat synthetic aperture radar imagery was taken to study the sea surface, but it has provided a new type of observational data to study geological features on land, particularly faults and other geological boundaries, surface textures, and soil moisture. Past climates are being inferred from paleontological data on fossil spores and pollens originally collected for biological studies of limited local areas.

The research community does not now have a coordinated effort to index these and other extant data sets and to distribute and update this index in the form of an electronic directory. Not knowing what data are available is clearly a barrier to international research in global change science. Efforts are currently under way to address this issue. For example, the International Geosphere-Biosphere Programme's Data and Information System has proposed identifying data sets at three levels: (1) by directories, identifying the existence of data sets; (2) by guides, containing information about their quality and other characteristics; and (3) by inventories, specifying the individual items that are present.[42]

Declassification of Environmental Data at the End of the Cold War

Observational data collected for military or espionage purposes are necessarily kept secret for some prescribed period of time, at least until the documented events, or the inherent evidence of the data collection techniques and technological capabilities themselves, can no longer compromise national security. Some of these data sets contain valuable historical data, particularly observations of certain locations or phenomena that are collected on a consistent, repetitive basis for many years, or even decades.

In recent years, both in the United States and especially in the countries of the former Soviet Union, the end of the Cold War has led to the release into the public domain of many of these data. An example is seismic data gathered for the purpose of underground nuclear test monitoring. Until the end of the Cold War, no regional recordings of Soviet nuclear explosions were available, nor were Soviet recordings of U.S. or other countries' test explosions. Now there is access to large amounts of these data by scientists outside the former Soviet Union. Other types of Soviet Earth science data, such as gravity and magnetics observations and Arctic oceanographic observations, also have been made available to the scientific community.[43] Likewise, U.S. data from some previously classified observational programs, including reconnaissance satellites[44] and undersea sensors,[45] have been made publicly available. The international availability of useful Earth science data has increased significantly with these data declassifications, and the committee encourages all governments to undertake similar reviews of classified retrospective data sets.

Improving International Access to Scientific Data in the Observational Environmental Sciences

A striking example of the benefits of extensive data collection and research for international management of environmental problems is evident in agreement by the nations of the world on a clear strategy for mitigating depletion of ozone in the stratosphere.[46] Not only was agreement reached in a limited period of time, but substitute substances and technologies also have been developed rapidly without a large economic impact on society.

Unfortunately, not all of the many global environmental and health problems can be confronted in the same way as was done for stratospheric ozone. In fact, many of the underlying research issues are extremely complex and interrelated. In the case of reducing the uncertainties regarding the much publicized global warming trend, extensive geophysical and biological data on the atmosphere, ocean, land surface, and cryosphere will be required on a global basis for long periods of time. The role of the ocean in the global carbon cycles and in the energetics of the atmosphere, the impacts of deforestation and desertification, the full implication of the radiatively active gases, and a host of interrelated natural processes need to be understood. Such understanding can be gained only by acquiring and analyzing comprehensive data sets on a global basis, with the active involvement of most, if not all, nations, and with the best efforts of the world's scientific community. But we do know from the stratospheric ozone problem that international agreement can be reached when adequate data and understanding of the problem are available to policymakers throughout the world. It is therefore essential that environmental and health data and information capable of describing our global atmosphere, ocean, and terrestrial system be fully and openly available. Moreover, making such basic data broadly available is fundamental to ascertaining the veracity and validity of the scientific process and of the resulting conclusions. If the data supporting the conclusions are not readily available to others for independent analysis, then the confidence in the research process and results will be undermined.

This is not to say, however, that all data must be made widely available as soon as they are generated. Indeed, an important reason for some period of delay is to ascertain the accuracy and integrity of the data and to prepare them for broader use, as discussed in the previous sections. The difficulties inherent in the collection and proper documentation of data by field researchers, or in the processing and organizing of large and complex data sets, can make a delay in the release of those data not only justified, but prudent. In addition, it is customary in many cases for a funding agency to provide the principal investigator or originating research group with the right to withhold public release of their data for a prescribed period of proprietary use, not only to adequately prepare the data for broader dissemination and use, but also to give the principal investigator an opportunity to analyze the data and to publish the first results. At the same time, there may be legitimate countervailing public policy reasons for early or even immediate availability of data, for example, data collected in publicly funded government programs such as meteorological satellite systems, which have both immediate operational and longer-term research applications.

While the availability of scientific data as soon as is reasonably possible should be the presumption, a single, uniform time period for the release of all data is neither sensible nor desirable. What is important is that the funding agency together with the community of scientists make a thorough evaluation of the competing interests guiding the release of its data prior to the initiation of every major data collection program, to establish the terms and conditions of data availability in consultation with the principal research and data user communities, and to subsequently enforce compliance. The 1996 NASA Science Policy Guide provides a good example of a data availability policy for nonproprietary scientific data obtained through public funds (see Box 3.4).

From a broader policy standpoint, the committee believes that the U.S. data management policy established for the U.S. Global Change Research Program in 1991 (commonly referred to as "the Bromley Principles" in reference to D. Allan Bromley, the President's Advisor for Science and Technology at that time) provides an excellent model for all major aspects of data availability and access. Box 3.5 contains the main points of that policy. The committee adapts the definition of "full and open exchange of data," subsequently developed by the NRC's Committee on Geophysical and Environmental Data,[47] as data and information are made available with as few restrictions as possible, on a nondiscriminatory basis, for no more than the cost of reproduction and distribution.

Unfortunately, the international exchange of data between research groups, government agencies, and scientific data centers, including the World Data Centers, is rapidly becoming more complicated, just at a time when full and open exchange is most needed to make progress on major global environmental problems. A growing number of government data centers outside the United States

**BOX 3.4
NASA's Data Availability Policy**

Ready access to data from NASA research programs and missions (via modern data archiving and communications technologies) by researchers not directly involved in the program increases the return on NASA research investments. **It is therefore NASA policy that nonproprietary scientific data obtained from NASA programs and missions will be made publicly available in usable form as quickly as possible.** (Nonproprietary data are data that may be distributed without violating patent, trade secret, or copyright laws or NASA's ability to obtain and protect U.S. government intellectual property rights.) Such data constitute a national resource that can be used by scientists, policymakers, and the public throughout the country to undertake new scientific studies, permit wider assessment of the validity of the results and conclusions from NASA missions, and facilitate broader public understanding of the value of NASA programs and missions.

The issue of data rights is a complex one that involves consideration of a wide range of competing factors including:

- the right of public access to data which has been obtained at public expense;
- the need to protect the original ideas which form the basis for competitively selected research (there is a strong tradition and body of law in the United States concerning the protection of intellectual property rights);
- the principle of fairness to investigators to allow them to pursue original ideas and hypotheses and to carry out the scientific investigations for which they were selected;
- the need to avoid the premature release of misleading results;
- the need to verify data prior to public release;
- the need for early release of data when such early release is critical to national needs or required for overarching public policy reasons;
- the need for early release of data for educational and public information purposes; and
- the need to protect data which may have a proprietary commercial application which may confer a competitive advantage, particularly to U.S. industries competing in the international marketplace.

A wide variety of approaches to data rights have been used:

- Virtually immediate release for scientific and public information purposes.
  — Shoemaker-Levy 9 observations
- Release of data as soon practical when data are considered reliable for general use.
  — Earth Observing System
- Restricted use of data during a limited calibration and verification period, after which verified data are deposited in a public archive.
  — Magellan
- Restricted use of data for a limited period of time (typically one year), after which verified data are deposited into an archive for general use.

> — Most past solar system exploration missions
> — Compton Gamma Ray Observatory
> — Hubble Space Telescope
> — Life and microgravity sciences research
> — In-space technology experiments (2 years)
> - Restricted use of data for an extended period to carry out investigations requiring the acquisition of data over a long interval of time; data are eventually archived.
> — Cosmic Background Explorer
> - Stringent restrictions placed on access to data on the basis of Privacy Act or other considerations.
> — Human research data
> — Proprietary data obtained through the use of NASA facilities such a wind tunnel test results
> — Restrictions that result from data purchase agreements
>
> **It is NASA policy that all nonproprietary scientific mission data be made publicly available after the shortest reasonable time in forms which permit a wide range of users to derive scientific, technical, and other benefits.** However, it appears that neither NASA nor the research community would be well served by the rigid adoption of a single uniform policy on the distribution and dissemination of data. Rather, the policy should be established for each mission or research program on a case-by-case basis. Well-understood and widely circulated criteria for making such determinations must be established. The approach to be taken for each program or mission will be spelled out in Announcements of Opportunity, Research Announcements, or other competitive mechanisms so that prospective participants understand the conditions of participation. Mechanisms may also have to be developed to assess adherence to NASA policies concerning the general availability of data. If a change is necessary in a previously agreed-to approach to data rights, NASA will consult the investigators affected to develop a mutually agreeable plan that meets the spirit of the principles set forth here.
>
> SOURCE: Reprinted from NASA Science Policy Guide (1996), available on-line at <http://dlt.gsfc.nasa.gov/cordova/guide.html>.

charge high prices for data and impose various dissemination and use restrictions. Ad hoc bilateral agreements between data centers and government science agencies are becoming commonplace. These agreements take on many forms and can lead to a situation in which individual scientists will no longer be able to obtain data for their projects without a major effort or large expense. Additional legal constraints on access to scientific data of all types are currently being implemented or considered, as discussed in depth in Chapter 5.

The committee thus recommends that internationally, in both intergovernmental and nongovernmental organizations, the full and open exchange of scientific data from publicly funded research be adopted as a fundamental principle.

> **BOX 3.5**
> **The "Bromley Principles" Regarding Full and**
> **Open Access to "Global Change" Data**
>
> The overall purpose of these policy statements is to facilitate full and open access to quality data for global change research. They were prepared in consonance with the goal of the U.S. Global Change Research Program and represent the U.S. government's position on access to global change research data.
>
> - The Global Change Research Program requires an early and continuing commitment to the establishment, maintenance, validation, description, accessibility, and distribution of high-quality, long-term data sets.
> - Full and open sharing of the full suite of global data sets for all global change researchers is a fundamental objective.
> - Preservation of all data needed for long-term global change research is required. For each and every global change data parameter, there should be at least one explicitly designated archive. Procedures and criteria for setting priorities for data acquisition, retention, and purging should be developed by participating agencies, both nationally and internationally. A clearinghouse process should be established to prevent the purging and loss of important data sets.
> - Data archives must include easily accessible information about the data holdings, including quality assessments, supporting ancillary information, and guidance and aids for locating and obtaining the data.
> - National and international standards should be used to the greatest extent possible for media and for processing and communication of global data sets.
> - Data should be provided at the lowest possible cost to global change researchers in the interest of full and open access to data. This cost should, as a first principle, be no more than the marginal cost of filling a specific user request. Agencies should act to streamline administrative arrangements for exchanging data among researchers.
> - For those programs in which selected principal investigators have initial periods of exclusive data use, data should be made openly available as soon as they become widely useful. In each case, the funding agency should explicitly define the duration of any exclusive use period.
>
> SOURCE: Data Management for Global Change Research Policy Statements, U.S. Global Change Research Program, July 1991.

The committee believes that such an agreement would significantly improve the ability of researchers to develop an adequate scientific understanding of our natural environment and the human condition, to address major problems facing the world community, and to broaden and enrich the knowledge base of all humanity.

Given that scientific data in all the disciplines—not just the observational sciences—are widely distributed globally in different archives or databases and that significant changes in data collection, storage, retrieval, and dissemination are steadily taking place, it is clear that some form of distributed data management strategy will be required to assure effective and efficient access by the scientific community. Considering the trends outlined in this chapter, and the imperative for broad and sustained international cooperation in environmental research, the committee concludes that the most viable and effective approach for the transnational flow of scientific data and information is through a system of connected international networks, each of which is the gateway to particular types of information. These data exchange networks, building on the successful models presented in Appendix C, would connect peer institutions for mutually beneficial rewards and collaborations, and provide data access to the research and education communities. The committee recommends the continued evolution of the existing distributed network of data centers as part of the global information infrastructure, with coordinated standards and procedures to provide unrestricted access at zero or low costs to data required for the study of regional and global problems.[48] This "network of networks" would provide connectivity to multiple data archives internationally and would serve as a coordinated source for important scientific data and information. Significant savings in research time, effort, and cost, as well as an overall enhancement of results, could be realized by using such a resource.

Terminology and Nomenclature in Biology and Related Fields

A significant barrier to sharing of information in the natural sciences is that subfields within disciplines have different languages, jargon, and usage. Without clear means for bridging resulting gaps in understanding, communication can be difficult. Moreover, lack of precision in terms themselves or in their use can lead to fundamental problems in searching for and interpreting data. A biologist, for example, may use the common name of an organism in recording and transmitting data without taking into consideration the limitations of the term or differences in usage. "Mouse" can signify any of a large number of small rodents. Peanuts are not nuts. Hospital records that indicate "atypical E. coli" without including the original observations justifying the label "atypical" do not communicate as much information as may be needed for treatment of a patient or for subsequent studies.

The complexity of what is being observed can also complicate precise description. Although the largest epidemiological studies describe far fewer events than does a short span of infrared satellite data, for example, the resulting biological database typically will have many more attributes associated with each event than are associated with, say, astronomical events. Likewise, a patient's hospital record often contains hundreds of different kinds of observations generated in the course of a diagnostic workup, and even a routine blood sample can yield data of at least a dozen kinds. Variability in the interpretation of each observation and associative reasoning play key roles in subsequently understanding the information.

A fundamental problem related to transfer of biological information in particular is the lack of a consultative body to standardize definitions for the words used to describe features of organisms. Dictionaries' coverage often is limited to narrow groups of organisms. There is no universal language of biology. The chemist knows with great precision what the term "sodium chloride" means; determining precise descriptors for aspects of biological entities presents challenges of another magnitude.

Consider, for instance, the term "spore" as used in a description of microorganisms. Spore-forming microorganisms consistently turn up in microbial biodiversity and bioconservation studies. But the taxonomy of these organisms is complex. Spore formation occurs in bacteria, fungi, protozoa, and algae, and the range of types of spores is also wide. Currently, the definitions and descriptions of spore types are very confusing but must be taken into account. Because of the lack of a comprehensive, authoritative treatment of the description of spores and spore types, many biodiversity studies are either in error or not understandable owing to misidentifications or confusing descriptions. Furthermore, the lack of a consistent vocabulary with respect to spores has legal and regulatory consequences. Intellectual property rights concerning strains of microorganisms require strict definitions and accurate, understandable descriptions of the properties of the strains, especially if a strain or its use is to be patented. Scientists, government regulators, patent officials, and industry lawyers, among others, all require agreed-upon definitions and standards for describing the various forms of spores and spore-related anatomical features of microorganisms.

The difficulties associated with describing and defining spores applies to many other descriptors used in biology. Further, as biologists adapt words from other disciplines and branches of biology, the meanings can drift. Terms developed by botanists to describe the shapes of leaves are also used in describing cell shapes in microorganisms, albeit with subtle or even major changes in meaning. "Obpyriform" signifies a pear-shaped leaf with the stem coming out of the blunt end. Applied to algae, the same term indicates pear-shaped cells with the blunt end in the forward direction in swimming. In addition, although "pear-shaped" assumes as a model the common commercial pear sold in most of the world, it is also the case that Asian "pears" have no obvious narrow end.

Problems of nomenclature extend to large-scale biological studies as well. For example, the lack of consistent classification schemes for land-cover vegetation and soils hampers international data exchange and can lead to errors of interpretation, especially by non-expert users. Current approaches to classifying land cover include the physiognomic, floristic, and ecosystematic approaches. It is generally agreed that existing land-cover maps cannot provide a globally consistent and up-to-date database for global change studies.[49] Satellite remote sensing is the only viable approach to developing a map of vegetation that is useful for global change research, and several satellite remote sensing activities now are addressing data requirements for characterizing land cover.[50]

For soils, as for land cover, there is no agreed-upon classification scheme. Soil is a three-dimensional dynamic entity whose properties—physical, chemical, and biological—vary dramatically in time and space. Observations of the causes and effects of these variations provide a valuable historical record of the components and processes that have produced current soil characteristics and conditions. With today's focus on research to understand human-induced changes and rates of change in the Earth system, the challenge to the soil science community is to provide a globally credible, compatible, and usable soil and terrain information system that can be integrated with information about other components of the Earth system. Only with open access to essential terrestrial information can intelligent decisions be derived about the Earth system and what, if any, human intervention is needed to protect it. Box 3.6 provides an overview of some of the nomenclature problems in this complex area.

The need for improved standardization of terminology, however, is not confined to biological or soil science databases. In any field, use of standardized terminology in a computerized database is vital in structuring the database, in digitizing data captured from printed sources, in accessing the database, and in interchange of data. The inflexibility and inexorable logic of the computer put new emphasis on control of terminology and on value-added system features such as thesauri, synonym files, and expanders for abbreviations and collectives terms. Without these aids, valuable information in databases may become inaccessible or incorrect, and incomplete or misleading information may be retrieved.[51]

Terminology-related barriers to understanding of data can be addressed only through internationally coordinated actions. Some efforts exist, such as the SOTER project described in Box 3.6. Another notable effort, a workshop to initiate mapping of the correspondence in terms associated with spores across all of microbiology, was supported with modest funding from the United Nations Environment Programme, the Committee on Data for Science and Technology (CODATA), and the U.S. National Science Foundation. The comprehensive *Dictionary of the Fungi*[52] is a widely accepted compendium of terms and their definitions. However, the *Systematized Nomenclature of Pathology*,[53] developed in the 1960s for computer entry of data, has not been widely adopted, either in the United States or in the rest of the world.

An important activity in the biological area has been CODATA's Commission on Standardized Terminology for Access to Biological Data Banks. The commission's encouragement led to the International Union of Pharmacology's establishment of a body to standardize the nomenclature for receptors for drugs. The International Committee on the Taxonomy of Viruses collaborates with the commission in its efforts to standardize the descriptors for viruses. The commission also is participating with the International Union of Biological Sciences and

BOX 3.6
Incompatible Soil and Terrain Information Systems

Today there is a critical need for detailed, universally compatible soil and terrain information for use in such applications as global change modeling, national resource planning and development, and plant breeding. Between 1960 and 1980 the Food and Agriculture Organization (FAO) of the United Nations, working with all nations, generated a Soil Map of the World, at a scale of 1:5,000,000, and published it jointly with UNESCO.[1] This major accomplishment provided a basis on which more current and detailed maps suitable for global change modeling might be built. A soil and terrain map at a scale of 1:1,000,000 is critically needed to study and model rates of change related to human activity in terrestrial ecosystems.[2]

For the industrialized countries, much more detailed soil maps exist at scales ranging from 1:10,000 to 1:500,000. However, for many of the less developed countries, the FAO-UNESCO Soil Map of the World may be the only soil map available for the whole country, although some countries may have detailed maps for some areas. Because of its lack of detail and the inadequacy of its original data, the use of the Soil Map of the World for national resource planning and development is often questionable. Unfortunately, FAO has no plans to produce a more detailed soil map of the world.

Since 1950, many agricultural and resource development programs have been implemented throughout the developing world, with technical and financial assistance from the United Nations and governmental and private agencies from industrialized countries. These assistance programs have included a number of localized soil mapping projects. As a result, in many of these countries soils have been mapped according to different soil classification systems. Often no attempt has been made to integrate one system with another. This situation has resulted in great confusion in many countries. The significant differences among classification systems confound the interpretation of these soil maps for national resource management, and make it difficult, if not impossible, to derive a credible national quality assessment of soil and land resources.

In 1986 the International Society of Soil Science embarked on an ambitious project to develop the World Soils and Terrain (SOTER) digital database at a scale of 1:1,000,000.[3] The first task was to develop a universal legend for describing both the cartographic units and the descriptive data for different soil and terrain categories. The second task was to develop a set of procedures that would make it possible to translate and correlate soil and terrain data (cartographic and descriptive) from any soil classification system to the universal SOTER database.[4] During 8 years of effort in several countries many improvements have been made in the SOTER procedures, but the number of constraints to progress remains formidable.

One of the greatest technical constraints is that there is no universally accepted soil classification system. Three of the major systems in use (i.e., systems that have been applied in relatively large areas) are the Soil Taxonomy system (a U.S. system developed with broad international collaboration), the French system, and the Russian system. These systems have been applied widely in specific projects in countries where the United States, France, and Russia have had collaborative development projects or strong ties. SOTER attempts to address this constraint by providing a universal legend under which any existing soil and terrain map of acceptable quality may be translated and correlated with the SOTER legend and entered into the SOTER database. In this process the data set for any country or classification system can retain its original identity.

Measurements of physical, chemical, and biological characteristics are essential for quantifying soil quality. One of the technical constraints in the measurement of soil properties is the lack of uniformity or standardization of analytical procedures. Another constraint on the comparison of soil properties from one classification system to another is that for many soil properties, different class limits are assigned for ranges of soil parameters used for classification purposes. Examples are the lack of uniformity in the classification of soil texture, that is, the amounts and distribution of different sizes of soil particles, and the absence of a universally accepted definition of slope—none-to-slight, slight, moderate, steep, very steep.

Another barrier to access to comparable soil and terrain data sets across international boundaries is the disparity in the conceptual use of quantitative data to delineate mappable soil differences. Some classification systems integrate much more quantitative analytic data into these delineations than do other systems.

Finally, there are intradisciplinary and interdisciplinary constraints. Within the applied discipline of soil science, there are many "subdisciplines," including soil physics, soil chemistry, soil mineralogy, soil microbiology, soil bioremediation, soil genesis, soil classification and survey, and soil degradation and reclamation. These groups often work in relative isolation from each other and may develop their own jargon, which may hinder access to data within the soil science community itself.

Efforts have been made in recent years to "connect" soil scientists with specialists in other disciplines, such as crop geneticists. Much more could be done, however, to remove the constraints to better use of soils information by plant breeders in the development of crops more suitable to prevailing soil conditions, such as cultivators of maize or rice that are tolerant (or resistant) to aluminum toxicity, which is prevalent in acid soils of the tropics. Many other scientists, land use planners, resource managers, civil engineers, environmental engineers, attorneys, resource economists, and others require specific kinds of information about soils and soil properties. Unfortunately, it continues to be very difficult for the non-soil scientist to effectively access and use the data.

---

[1]Food and Agriculture Organization (1980), *FAO/UNESCO Soil Map of the World*, Food and Agriculture Organization of the United Nations, Rome.

[2]M.F. Baumgardner (1993), "The Critical Need for a World Soils Database for Global Change Modeling," in *Proceedings of International Workshop on Soils and Global Modeling*, International Geosphere-Biosphere Programme, Stockholm.

[3]International Society of Soil Science (1986), "World Soils and Terrain Digital Databases at a Scale of 1:1M (SOTER)," project proposal, M.F. Baumgardner (ed.), ISSS, Wageningen, the Netherlands.

[4]International Soil Reference and Information Centre (1993), *Global and National Soils and Terrain Digital Databases: Procedures Manual*, ISRIC, Wageningen, the Netherlands.

the International Union of Microbiological Sciences to establish the System 2000 network to assemble a database of scientific names of all biota.

All these efforts, however, represent a small part of what needs to be done to make data and results more accessible. The mechanisms for establishing standards for words, formats, and storage and retrieval conventions in biological information management need to be improved. To be effective and accepted, this must be done on a truly global basis. In general, standardization of terms in science today is either carried out or validated by the appropriate ICSU body. For example, the CODATA Task Group on Fundamental Constants is the recognized authority in nomenclature for fundamental constants. In biology, the Codes of Nomenclature are promulgated by ICSU components appropriate to the discipline.

The committee suggests that the CODATA Commission on Standardized Terminology for Access to Biological Data Banks be enhanced into a true consultative body for this purpose. The commission would need funds sufficient to provide effective standard-setting services to the biological community. Expansion of personnel and increased collaboration with other ICSU and outside scientific organizations would be necessary for both functional and political reasons. This should be an ICSU function, coordinated by CODATA, because there is no other established international source of such standard setting in the biological sciences.

Data Compatibility in the Laboratory Physical Sciences

The barriers to the international exchange of scientific data in the laboratory sciences generally are not as complex as those in the observational sciences, partly because of the difference in the volumes of data accumulated and used in day-to-day research and partly because of the ways in which the disciplines have evolved. In the physical and the laboratory biological sciences, for example, full compilations of data have always been published in textbooks and in articles in professional journals available throughout the world, whereas the data of the observational sciences in many cases have been accumulated only in government records.

Barriers to international data exchange in the laboratory sciences concern ease of access to data and the use of those data. Today, the effective exchange of virtually all scientific data requires that they be in electronic format. For manuscripts, exchange can be straightforward if scientists adopt a common word-processing language such as TeX or LaTeX, now used worldwide in many communities in the physical sciences. Scientists need to be able both to generate a computer-readable manuscript that can be decoded and read on all computer platforms and to decode and read whatever other scientists may similarly provide. Establishing a common set of tools to ensure such compatibility can be difficult, more so for simulations and animations than for text. Because of the volume of data they involve, simulations and animations need to be compressed for storage and transmission, a problem that should be easy to resolve since they invariably originate in computer format. However, making them accessible to arbitrary platforms requires either considerable sophistication or standardization, or both.

Converting databases created in hard copy to electronic format can be a costly enterprise, but is nevertheless far cheaper than erecting library buildings. Considerable care is needed to ensure that the original data are not compromised in the process of generating the electronic version.[54] In recent years, most data transferred automatically from paper to computer have been captured and stored as images of the printed pages. The alternative is to store the data as text, apart from components that are true images, such as molecular structures; the increasing availability of optical character-reading software is making textual storage practical and economical. The large number of databases in the physical sciences that have been developed by the National Institute of Standards and Technology (NIST) have traditionally been available only in relatively expensive, hard-copy books. NIST plans to provide on-line access to all data it collects and compiles and recently has developed on-line access, with search capabilities, to databases critical to research in chemistry and physics.[55] There is a great difference between data stored electronically as images, which cannot be manipulated by the user, and data stored as digitized alphanumeric information, which can be treated as normal text and tables. Data in most modern databases in the physical sciences are generally not static, especially when the databases are stored electronically and hence can be updated as the information improves. Thus there is always a continuing responsibility and expense to maintain and disseminate those databases as they evolve. If the data are stored only as images, such maintenance is difficult and costly. Storage, maintenance, and distribution become vastly easier and more efficient if the information is in the form of a true relational database, in alphanumerics, with user-friendly search capabilities, qualities that require expense and technical sophistication to implement.

In some areas of the physical sciences, notably materials science and chemistry, the fragmentation of the data into numerous, autonomous, and often incompatible databases continues to be a considerable barrier to access. Many small data files exist, often maintained by individuals, with a plethora of formats and a range of quality levels. When there are several databases, many means of access to them, and inadequate directories to locate and search them, it is difficult to know what information a particular system of databases includes, how to locate sources for information that they do not cover, and how to assess the quality of the data. The problem is further exacerbated when some data are in journal form, others in hard-copy manuals, and still others in a variety of electronic databases, each of which may be on a different platform, often with limited search capabilities. This is in contrast to situations in areas such as atomic and nuclear physics, in which data have traditionally been compiled and disseminated from a single source, or at least in a standardized format. The dissemination of materials science and chemistry databases remains fragmented, and the broad range of researchers in these fields still need better access to them.

Most of the data in the databases of the physical sciences are needed to carry on the basic research that the funding agencies support. Just as the funding agencies support the hardware necessary to do research, these agencies also carry a responsibility to support the data components of the infrastructure necessary to conduct research. Also, just as support for basic research needs to be protected because of the likelihood of long time intervals between the conduct of the research and its eventual applications, so should the development and maintenance of databases be protected from short-term fluctuations in budgets or varying needs for the data in industrial applications. The development of databases includes the compilation and evaluation of data from the variety of sources of the data. Once developed, it is critical that databases be maintained and continuously updated as new, relevant data become available. The dissemination should be via a variety of platforms and should be in user-friendly forms, with cross-referencing to files maintained by other agencies, or available via other electronic media.

The committee believes that science agencies should maintain responsibility either directly or under subcontract for the development, management, retention, and dissemination of electronic databases that are the product of their research programs. Within the United States, the Office of Science and Technology Policy should develop an overall policy for the long-term retention of scientific data, including a contingency plan for protecting those data that may become threatened with the loss of their institutional home.[56]

ACCESS TO SCIENTIFIC DATA IN DEVELOPING COUNTRIES

The international exchange of scientific data has a scope beyond that of the large scientific communities in the technically and economically developed parts of the world. While much of this report reflects the research atmosphere in which its contributors work, it is especially important to address aspects of the subject associated with disparities of wealth and resources among nations, the cultural differences with which nations address their societal problems, and the varying ways nations assign their priorities.

The differences in priorities are especially marked in the spectrum of ways in which nations, from one end to the other of the scale of development, consider scientific and technical matters. More industrialized and wealthier nations choose to invest discretionary public funds in basic sciences, such as high-energy physics and astronomy, as well as in applied and developmental science and technology. Nations toward the other end of the developmental scale put little emphasis on sciences with long-term public-good payback and put most of the resources they have into applied sciences such as agriculture, aquaculture, medicine, and, recently, biotechnology.

As a consequence of the availability of discretionary resources, scientists in more industrialized nations traditionally have been able to obtain reliable, up-to-date research equipment, computers, communications infrastructure, and information resources. The scientific communities in developing countries have not had such advantages. In the context of this report, this has meant that the scientists in developed nations have had much better access to data and to the underlying means of communication than their colleagues in other nations, who consequently have not been able to take full advantage of their talents.

One of the great challenges in the advancement of science that now faces the international community is to use electronic acquisition, management, storage, and distribution of scientific data to reduce the gap between those who have had easy access to the fruits of scientific progress and those who have not. Because of the decreasing costs of electronic technology, compared with the rising costs of traditional means of storing and transmitting scientific data, the opportunity is now opening to make advances in bringing scientists in developing countries much more deeply into the circle of their colleagues in developed countries. There will be problems and outright barriers to confront in the process of reducing this gap, but the situation now offers brighter possibilities than at any time since science became a major, worldwide enterprise.

In this section, the committee reviews some of the issues in data access in the context of this asymmetrical relationship between the developed and developing world. The constraints to data access within developing countries are considered first. These include both the limited capability to generate new scientific data and the problems facing indigenous scientists who want ready access to data from outside sources. Such limitations lead to underutilization of the talents of those scientists because they cannot easily stay abreast of advances in their fields. The committee then examines the ability of scientists in developed countries to obtain useful data based on work in developing countries.

Constraints on Data Access Within Developing Countries

Basic to any consideration of constraints on access to data is the economic situation in developing nations. Economic limitations on access to scientific data are manifest primarily in the inadequacy of communications infrastructure and related research equipment (pointed out in Chapter 2), as well as in insufficient resources for training and education. Another important set of constraints not as deeply affected by the lack of resources is organizational inadequacies.

In many developing countries today, gaining and maintaining access to international sources of scientific data and literature are very difficult. University libraries and research institutes in these countries cannot afford to subscribe to the major scientific journals, publications that tend to be readily available to scientists in wealthy nations. Databases that are available even at very low rates, such as the marginal cost of reproduction, can be prohibitively expensive. Contact and sharing by scientists in non-industrial nations with scientific colleagues in other countries can be extremely limited. Although there have been some notable efforts on the part of organizations such as the American Association for the Advancement of Science and UNESCO to provide scientists in developing countries with printed copies of scientific data and information,[57] much more could and should be done to improve such sharing of information. The committee therefore recommends that until affordable and ubiquitous electronic network services are available, national and international scientific societies and foreign aid agencies should establish or improve their existing efforts to send extra stocks of scientific publications to libraries and research institutions in developing countries that need them.

Training and Education Considerations

The governments of most countries recognize that education, particularly higher education, is vital for the creation of a solid national base for scientific endeavors and economic growth. The poorest nations typically send their students abroad for advanced education and specialized training, often in applied disciplines deemed most useful upon the students' return. Following completion of their postgraduate education and research abroad, however, a large number of these highly skilled scientists do not return to their home countries, effectively negating for the home country the immediate broader benefits of their training.[58] Many of these countries cannot provide a sufficiently supportive environment, including the necessary research infrastructure and funding, to attract and keep scientists. Further, lack of ready access to current information leads to professional obsolescence. The "brain drain" from the poorer to the wealthier nations is a serious constraint to the generation of new knowledge in the developing countries.

In addition to the limitations of the available data management and communications technology, training in the use of available technology is limited as well. The growing sophistication of both hardware and software tends to make their use more efficient and eases the training burden in some ways, but it increases it in others. Basic functions of the computer system are becoming increasingly automated. However, the functional power of the systems increases the demand for and use of more complicated techniques for management, analysis, and dissemination. An important related problem is a lack of adequately trained personnel for servicing such complex equipment.

At the most basic level is the lack of instructional support for the neophyte computer user. For example, in courses taught under United Nations auspices on topics such as use of computers in microbiology, the students in developing nations overwhelmingly request supplemental training in the use of computers for data management and analysis, and in on-line access to data and information resources. Generally speaking, much more instructional outreach in basic computer data management and communication skills is needed.[59]

The committee recommends that international development organizations, together with professional societies, provide targeted training programs for scientists in the use of computers, with emphasis on the management of digital data in specific disciplines.

Organizational Issues

There are many organizations that provide bilateral and multilateral assistance to scientists in developing countries, although few are focused primarily or exclusively on scientific data issues. These organizations support scientists through a variety of mechanisms. Some provide scientific data and services directly to researchers in developing countries, others provide access to data through journal subscriptions and travel grants to international scientific conferences, some provide Internet connections and information technology services, and others promote and provide training and education. Examples of national and international government institutions, nongovernmental organizations, not-for-profit organizations, professional societies, and private-sector firms that provide these types of services are described briefly below:[60]

- *U.S. government.* Within the United States, the federal government, primarily through the U.S. Agency for International Development (USAID), provides foreign assistance for activities of scientists and engineers in less developed nations.[61] Other federal agencies such as NASA and the Department of Agriculture assist scientists by providing data resources and data management services.[62] Finally, the Department of State, through its Bureau on Oceans, Environment, and Science, indirectly provides assistance through negotiating and monitoring environmental agreements and conventions that have significant cooperative research and data exchange provisions.[63]
- *Intergovernmental organizations.* Many intergovernmental organizations provide assistance to scientists and researchers in developing countries by providing data and information, training and education, and assistance with information technology. The lead player in this arena is the United Nations, primarily through the United Nations Educational, Scientific, and Cultural Organization (UNESCO), United Nations Development Programme (UNDP), United Nations Environment Programme (UNEP), Food and Agriculture Organization (FAO), World Health Organization (WHO), World Meteorological Organization (WMO), and the World Bank.[4]

Regional organizations, such as the Organization of American States (OAS)[65] and the Pan American Health Organization, also promote science and technology in developing countries through regional activities. The European Community pursues scientific and technological cooperation with developing countries as well, particularly with the aim of generating knowledge and technologies needed to help achieve sustainable development.[66]

Finally, various ad hoc intergovernmental groups and committees have been organized to coordinate activities related to major international research programs as discussed above in this chapter. Many of these groups have subgroups devoted to different data management issues, including activities focused on developing countries. For example, the Committee on Earth Observation Satellites (CEOS) coordinates all spaceborne Earth observation missions among the spacefaring nations. CEOS has established a "Plan of Action for Support to Developing Country Activities by CEOS Participants."[67] Box 3.7 presents a number of useful "lessons learned" by CEOS participants in providing support to developing countries.

- *Nongovernmental organizations (NGOs)*. International NGOs, such as the Third World Academy of Sciences,[68] the International Council of Scientific Unions,[69] and the International Foundation for Science,[70] collaborate with U.N. programs and agencies to provide scientific and technological support to developing countries.

The Consortium for International Earth Sciences Information Network (CIESIN) is an example of a national NGO, with broad international scope, that provides data and services to scientists in the developing countries. In addition to providing "global and regional network development, science data management, decision support, and training, education, and technical consultation services," CIESIN is the World Data Center A for Human Interactions with the Environment.[71]

Many national and international not-for-profit organizations also assist scientists in developing countries via different mechanisms. The Sabre Foundation's Scientific Assistance Project provides educational materials in the form of books and journal subscriptions and an Internet-based technical assistance program to institutions and individuals in the former Soviet Union and Eastern Europe.[72] The International Science Foundation was established by George Soros in 1992 to assist scientists in the former Soviet Union and the Baltic States by promoting contacts with the international scientific community, providing access to scientific data and information, and establishing international communications links.[73] The International Research and Exchange (IREX) Board promotes academic exchanges between the United States and the former Soviet Union and provides professional training, technical assistance, and policy programs.[74] Other organizations, such as the Volunteers in Technical Assistance (VITA), contribute information services and technology to developing countries to improve their quality of life.[75]

National and international scientific and engineering societies and associations play an important role as well. For example, in addition to the African libraries program described above, the American Association for the Advancement of Science has promoted regional collaborations between scientists in developing countries.[76] Some professional organizations provide travel grants to allow individual scientists from developing countries to attend international sci-

BOX 3.7
CEOS "Lessons Learned" Regarding Support to Developing Countries

The Committee on Earth Observation Satellites (CEOS) compiled the following list of principles based on the experiences of its members in providing technical assistance to developing nations:

- Development projects should be planned in partnership between donors and local institutions in response to real needs of in-country decision-makers. Decision-makers need to be convinced of the utility of such activities in order to create the appropriate environment for sustainable operation.
- Projects supported by joint efforts of space agencies with development assistance organizations can benefit from combining important skills in both science/technology and sustainable development.
- Pilot projects should be selected with their later operational requirements in mind. To be considered successful, a pilot project will provide the foundation for ongoing routine application of the demonstrated capability. This suggests the use of affordable technology and readily available data. Projects aimed at improving indigenous capability to perform already ongoing operations are more likely to succeed.
- Documentation prepared for use by developing country users should be available in a language readily understood locally, using minimal technical jargon, to be easily understandable by the target audience.
- Data and information for developing countries should be on media appropriate for the users, avoiding electronic formats requiring equipment that is not available. Easily reproducible text and imagery will often be more readily usable than sophisticated digital products. At the same time, consideration should be given to improving local infrastructure so that media such as CD-ROMs can progressively be used in developing country applications.
- Expertise in developing countries must not only be created but also be sustained. This suggests holding local training courses and emphasizing "training the trainer." Improving existing educational institutions rather than creating new training centers can enhance the sustainability of the educational process.
- Local reception of satellite data can be an effective tool in identifying practical applications and demonstrating the value of the data in the local environment. Equipment installed in developing countries must be designed to be rugged and easily maintained with locally available capabilities.
- Satellite data alone will not contribute to development unless it is transformed into useful information and disseminated.
- Countries that have successfully applied satellite technology to development problems can serve as examples and share their experience and expertise within their regions and more broadly. Their experience may be more relevant and more applicable to other developing country situations than the approaches used in industrialized countries.
- Development assistance projects should be structured with sufficient flexibility to respond to the unexpected events that often occur.

SOURCE: CEOS; available on-line at <http://gds.esrin.esa.it:80/0xc06afc3d_0x000291f2;internal$sk=041858E7>.

entific and technical conferences. Other efforts include the American Society for Mechanical Engineers' partnership with the Mechanical Engineering Research Institute of the Russian Academy of Sciences to promote the application of environmental and energy-related technologies to establish a technology transfer mechanism between the two organizations.

- *Private sector.* A number of private sector organizations also provide assistance to scientists in developing countries. This assistance is usually indirect, through the financial support of international NGOs such as the Third World Academy of Sciences and ICSU.

Many of the problems cited above in this section are exacerbated by a lack of effective organizational structures or institutional mechanisms for involving scientists within developing countries in the decision-making process regarding scientific research, much less data access issues. However, foreign aid agencies in the developed countries and intergovernmental development organizations are known not to involve scientists in their decision-making process either. For example, U.N. funding agencies respond almost exclusively to requests from the foreign ministries of member countries. The foreign ministries in developing countries almost never utilize scientists in decisions. The result is a dearth of funding applications for scientific infrastructure capacity building, which is essential not only to support indigenous scientific research efforts, but also to encourage economic development. An analogous situation is evolving in USAID, where science once flourished, but where the involvement of scientists in internal planning and funding decisions is eroding rapidly.

Of course, some success stories do exist. For example, Vietnam, concerned about environmental pollution as well as the need to build biotechnology capacity, arranged for scientists at many levels to collaborate in developing a national plan in microbiology and biotechnology infrastructure capacity building for submission to the Global Environment Facility of the U.N. through the United Nations Industrial Development Organization.

With regard to improving access to scientific data in developing countries, the committee makes the following recommendations:

- Scientists in developing countries should be encouraged to organize to promote the policy of full and open access to scientific data in their own countries, as well as to make their data available internationally.
- Foreign aid agencies should (i) make available to individual scientists in developing countries more direct, peer-reviewed grants that include support for access to data, and (ii) facilitate the involvement of scientists in such nations in their own countries' capacity-building initiatives, research policy decisions, and national database construction efforts.

Constraints on Access to Data from Developing Countries

The constraints caused by inequities among nations in access to scientific data are felt most severely in those sciences concerned with inherently international issues, such as food production, biodiversity, the prevention and cure of communicable diseases, global climate change, and other Earth system processes. Each of these areas of concern requires international research and approaches to problem solving. As discussed above in this chapter, developing this essential understanding requires the generation of globally compatible, accessible, and usable data sets related to terrestrial ecosystems, the physical environment, and human activities. Collaboration of the scientific community in every nation, rich and poor, in the generation of global observational data sets and the subsequent full and open transnational flow of those data is imperative; its need cannot be emphasized too strongly.

For example, in the Earth and environmental sciences, particularly in global change research, it is essential to integrate remote sensing data with "ground truth" in situ observational data in the creation of consistent and valid data sets. Without this integration, the value of the data products and research results can be undermined considerably. The in situ data are generated by individual workers and organizations in many different countries. Maintaining cooperative activities through which the in situ data are reliably supplied is essential for the success of international research projects. Many of the gaps in the collection and dissemination of in situ data occur in developing countries, where a lack of resources and other barriers make such cooperative activities difficult.

As this report documents, the more wealthy industrialized nations have developed a broad range of international research initiatives, largely supported by a policy of full and open access to scientific data. Although significant problems remain and new barriers to effective collaboration continue to arise, there are sufficient incentives and resources for sharing of data by scientists in the developed world. However, the sharing of scientific data—particularly data for fundamental research—tends to be a much lower priority for many of the less wealthy, nonindustrialized nations. Success in ensuring full and open transnational flow of scientific data among these nations may depend on the degree to which the industrialized nations are responsive to the needs of the nonindustrialized nations and can provide incentives for their participation. An example of such a need involves the inability of developing nations to pay for large-scale disease treatment programs. Box 3.8 describes the mutual benefits possible in collaborative data sharing.

As discussed above, incentives for participation by developing nations might include assistance in the development of human resources, the provision of equipment, and the general improvement of the research and communications infrastructure. Significant efforts should be made by the scientific community of the wealthier nations to include scientists from the nonindustrialized nations as part-

> **BOX 3.8**
> **Examples of Successful Transnational Data Collaborations**
>
> The World Data Centers (WDCs) have sponsored several "data rescue" projects in developing countries. In some cases, modest funds and sometimes equipment have been provided to help local scientists digitize older time-series data as part of the effort to build the International Geosphere-Biosphere Programme's and other studies' global digital databases to document trends and changes. Local scientists thus get their own data back in digital form on diskettes or CD-ROMs, depending on the technology they have. These efforts further not only their own work, but also the work of global change researchers internationally. Also, the WDC-A Oceanography (NODC/NOAA) data rescue project provides the opportunity for local research groups to help to produce for the first time historical and highly useful analyses and maps of global ocean climatic changes.

ners in the global scientific enterprise, particularly in research initiatives that are fundamentally dependent on the availability of global data sets or in studies addressing basic needs such as disease control and prevention.

For example, some attention is being paid to the searching of genomic information useful for preventing tropical diseases,[77] and some of this research is being carried out in developed nations. However, greater emphasis on understanding such diseases would follow from enhancement of the infrastructure for expertise in biology and biotechnology in developing nations. Developed countries' promotion of such advances would not be purely altruistic. Leishmaniasis, a disease usually associated with the tropics, infected troops in Desert Storm and is present on both sides of the Texas-Mexico border. Tuberculosis usually is not perceived as a tropical disease, per se, yet resistant strains of the bacteria from the tropics have found their way into populations in the developed nations. The motivation for trying to identify and locate genes possibly conferring resistance in populations where the diseases are common would be deepened by researchers' proximity to and familiarity with the effects of such diseases, if the resources and personnel were available in the affected countries. Economic constraints might also be lessened for studies in developing nations, where labor costs, even for highly trained research scientists, are much lower.

Another barrier to the collection of data is evident in field studies related to biodiversity in developing countries. It is well known that the greatest concentrations of the planet's biodiversity occur in developing countries. However, the resources to study and exploit the diverse gene pools for biotechnology lie largely in the developed nations. In this area, as in all of science and technology, professionals in developing countries generally lack access to all the data and information needed to support their work. Further, considerations regarding intellectual property are more complicated in biology than in most other disciplines, because the biological materials themselves are repositories for scientific information. For this reason, bioprospecting for new gene pools in tropical countries by commercial and other interests from industrialized nations has become a contentious issue on a global scale.[78] For example, Brazil will no longer allow the sampling of biota by non-Brazilians and will not allow export of biota.[79] In such cases, the study of these materials is limited to what the country can do with local resources. Data that are produced in this way are sequestered rather than shared with the general scientific community. Other unanticipated problems can arise in this context as well, as Boxes 3.9 and 3.10 illustrate.

With regard to in situ data collection efforts in developing countries, the committee recommends the following actions: the ICSU, together with funding agencies and nongovernmental bodies, should strengthen its efforts to assist developing countries in undertaking their own scientific studies and encourage scientists engaged in such studies to take active roles in the international scientific community, where their efforts can be appreciated and used. Legal and procedural protocols must be developed to provide for fair and equitable sharing of any resulting intellectual property. This would not only help create indigenous

---

**BOX 3.9
A Hobson's Choice**

The following example of a trade-off between two unpalatable options was provided by the Consortium for International Earth Sciences Information Network (CIESIN) in response to the committee's "Inquiry to Interested Parties":

One unexpected experience is in the balancing of data access privileges with the access of researchers to pursue their research in specific countries. Our experience includes an instance where a multi-year program to collect and integrate socioeconomic and environmental data in an African country was successfully completed, the data conveyed to CIESIN for sustaining access, then the government of the subject nation was ousted through a violent and protracted coup. The successor government did not agree with the predecessor government, in terms of allowing open access to those data collected and provided by its agencies to the CIESIN-sponsored researchers. Thus, they wanted to prohibit future release of data already out of their physical custody.

The clear implication was that failure to comply with these newly implemented restrictions would cause further restrictions of follow-on research projects of the type CIESIN initially supported with UNEP [United Nations Environment Programme] and others. The trade off between restricting data access and restricting research access for future collection is an unsavory and unforeseen challenge that is likely to recur in that region and elsewhere, as political instability ensues. Future governments may decline to honor the information sharing policies of their country. This dilemma threatens the free and open access of data on a sustaining basis and raises significant questions about where the locus of ownership of data is after governments are replaced, peacefully or through violent actions.

> **BOX 3.10**
> **Can Data Be Too Accurate?**
>
> The following is an excerpt from a message that is part of a discussion on the Internet list server, <biodiv-1@bdt.org.br>. This discussion group emphasizes global biodiversity, conservation of habitat and biota, and information regarding these areas. The author of this message is Jeff Waldon. The message illustrates an important but little appreciated aspect of the tension between free dissemination of information and commercial and nonscientific private interests:
>
>> The debate is whether release or restriction of sensitive locational information is the best thing for conservation. There are cases of collectors using such information to decimate rare and endangered species at a site (e.g., the recent arrest of butterfly poachers that targeted National Parks in the western United States). On the other hand, there are other examples of species protection because the landowner was informed of the existence of a rare animal or plant. I have been involved in the development of information systems for about 10 years, and I have heard both sides argued strenuously. My personal feeling is that the "boogie man" collector is real, but in most cases we overreact to his presence. We are losing many more populations of threatened and endangered species because of ignorance rather than malice.
>>
>> We have developed a compromise in our systems whereby we release sensitive information on species, but the locational data accuracy is reduced to help reduce the likelihood that a collector might successfully collect at that site. If a development project for the public is reviewed, and more accurate information is required, that information is provided at the discretion of the biologist working with the requester. I come from the academic school of thought that relies on the free interchange of information, and this compromise strikes me as still too restrictive at times. On the other hand, government employees are bound by laws and policies that make them accountable for their actions including the consequences of releasing information on the location of threatened and endangered species, and I see their dilemma.
>
> SOURCE: Jeff Waldon, personal communication, 1995, used with permission.

data resources and promote a greater interest within nations of the developing world in obtaining a more thorough understanding of their own resources, but also lead to more fruitful international cooperative research.

RECOMMENDATIONS ON DATA ISSUES IN THE NATURAL SCIENCES

The recommendations set forth below are addressed to all individuals and organizations with responsibilities for managing scientific data acquired with public funds.

1. Governmental science agencies and intergovernmental organizations

SCIENTIFIC ISSUES   101

should adopt as a fundamental operating principle the full and open exchange of scientific data. By "full and open exchange" the committee means that the data and information derived from publicly funded research are made available with as few restrictions as possible, on a nondiscriminatory basis, for no more than the cost of reproduction and distribution.

2. The International Council of Scientific Unions (ICSU), together with the scientific Specialized Agencies of the United Nations, the Organisation for Economic Co-operation and Development Megascience Forum, and the national science agencies and professional societies of member countries, should consider developing a distributed international network of data centers. Such a network should draw on the strengths of successful examples of international data exchange activities as described in Appendix C of this report, including, in particular, the ICSU World Data Centers, and become a prominent part of the global information infrastructure that has been proposed by the "Group of Seven" nations. To facilitate the international dissemination and interdisciplinary use of scientific data, all public scientific data activities, including the network of data centers, should plan for and commit to providing the human and financial resources sufficient for carrying out the following functions:

    a. Involve experts from the relevant disciplines, together with information resource managers and technical specialists, in the active management and preservation of the data;

    b. Develop and maintain up-to-date, comprehensive, on-line directories of data sources and protocols for access;

    c. Provide documentation (metadata) adequate to ensure that each data set can be properly used and understood, with special attention given to making the data usable by individuals outside the core discipline area. This problem is particularly acute within the biological sciences, in which imprecision and variations in taxonomic definitions and nomenclature pose significant barriers to communication, even among the biological subdisciplines. The committee suggests that the CODATA Commission on Standardized Terminology for Access to Biological Data Banks be enhanced into a true international consultative body and that similar mechanisms be developed for other disciplines, as needed;

    d. Incorporate advances in technology to facilitate access to and use of scientific data, while overcoming incompatibilities in formats, media, and other technical attributes through vigorous coordination and standardization efforts;

    e. Institute effective programs of quality control and peer review of data sets; and

    f. Digitize all key historical data sets and ensure that every important condition for the long-term retention of data be met, including the adoption of appropriate retention and purging criteria and the timely transfer of all data sets to new media to prevent their deterioration or obsolescence.

3. The ICSU and other professional scientific societies should encourage the study of, and publication of peer-reviewed papers on, effective data management and preservation practices, as well as promote the teaching of those practices in all institutions of higher learning.

4. All scientists conducting publicly funded research should make their data available immediately, or following a reasonable period of time for proprietary use. The maximum length of any proprietary period should be expressly established by the particular scientific communities, and compliance should be monitored subsequently by the funding agency.

5. As a corollary to recommendation 2.a above, publicly funded scientific databases should be maintained either directly or under subcontract by the government science agencies with the requisite discipline mission and need. In the United States, the Office of Science and Technology Policy should develop an overall policy for the long-term retention of scientific data, including a contingency plan for protecting those data that may become threatened with the loss of their institutional home.[80]

6. With regard to improving access to scientific data in developing countries, the committee makes the following recommendations:
   a. International development organizations, together with professional societies, should provide targeted training programs for scientists in the use of computers, with emphasis on the management of digital data in specific disciplines.
   b. Foreign aid agencies should (i) make available to individual scientists in developing countries more direct, peer-reviewed grants that include support for access to data, and (ii) facilitate the involvement of scientists in such nations in their own countries' capacity-building initiatives, research policy decisions, and national database construction efforts.
   c. Scientists in developing countries should be encouraged to organize to promote the policy of full and open access to scientific data in their own countries, as well as to make their data available internationally.
   d. The ICSU, together with funding agencies and nongovernmental bodies, should strengthen its efforts to assist developing countries in undertaking their own scientific studies and encourage scientists engaged in such studies to take active roles in the international scientific community, where their efforts can be appreciated and used. Legal and procedural protocols must be developed to provide for fair and equitable sharing of any resulting intellectual property.

e. Until affordable and ubiquitous electronic network services are available, national and international scientific societies and foreign aid agencies should establish or improve their existing efforts to send extra stocks of scientific publications to libraries and research institutions in developing countries that need them.

7. Finally, the ICSU, together with the principal national and international scientific organizations mentioned in Recommendation 2 above, should convene a series of major international meetings to initiate meaningful action on these recommendations.

NOTES

1. Privacy issues, which become especially important in the social sciences and clinical research, were judged to be of only tertiary concern in the context of most of the disciplines examined in this study, and thus are not addressed in any detail in this report.
2. In some areas of the experimental sciences, it is standard practice for researchers to publish general results, such as structures of protein molecules, but retain details, such as precise coordinates of the atoms, for some limited period of time, during which they may pursue the implications of their own measurements. In many instances, particularly in the observational sciences, principal investigators are allowed to keep data sets proprietary for some specified period of time in order to be able to analyze them and publish their results first. This issue is discussed below in this chapter.
3. National Research Council (1995), *Preserving Scientific Data on Our Physical Universe: A New Strategy for Archiving the Nation's Scientific Information Resources,* National Academy Press, Washington, D.C.
4. Cosmic-ray research is an exception here. While it is based largely on observations rather than experiments, it has been classified traditionally in physics, rather than astronomy or space science. It overlaps all of these, of course.
5. For a more detailed discussion of the differences between experimental and observational data, see National Research Council (1995), *Preserving Scientific Data,* note 3.
6. For a comprehensive listing of most internationally available data sets from space missions, see the NASA Goddard Space Flight Center's National Space Science Data Center home page at <http://nssdc.gsfc.nasa.gov/ >.
7. For a broad listing of international WWW servers covering all aspects of Earth science data and information, see the NASA Global Change Master Directory at <http://gcmd.gsfc.nasa.gov/cgi-bin/pointers/>; see also <http://gds.esrin.esa.it:80/>.
8. A. Maddison (1995), *Monitoring the World Economy: 1820-1992,* OECD, Paris, 255 pp.
9. T.F. Malone (1995), "Reflections on the Human Prospect," in *Annual Review of Energy and the Environment* (R.H. Socolow, ed.) 20:1-29, Annual Reviews, Palo Alto, California.
10. See <http://www.usgcrp.gov > for additional information on the U.S. Global Change Research Program and related data activities, and <http://www.igbp.kva.se/index.html> for information on the International Geosphere-Biosphere Programme.
11. See the WWW Virtual Library for a comprehensive index of biological data and information at <http://golgi.harvard.edu/biopages>. See also a listing of sources of international biological information on the Internet on the Web site of the U.S. Geological Survey's Biological Resources Division at <http://www.its.nbs.gov/nbii/iao/ibii.html>; and the Biotechnology Indus-

try Organization's compilation of biotechnology databases at <http://www.bio.org/educ/dbasef.html>.
12. See, for example, National Research Council (1996), *Statistical Challenges and Possible Approaches in the Analysis of Massive Data Sets*, National Academy Press, Washington, D.C.
13. Genevieve J. Knezo, (1994), "Major Science and Technology Programs: Megaprojects and Presidential Initiatives, Trends Through the FY 1995 Request," Congressional Research Service, Washington, D.C., March 29, p. 1.
14. Congressional Budget Office, July (1991), "Large Non-Defense R&D Projects in the Budget: 1980-1986," CBO, Washington, D.C. Unfortunately, more recent statistics are not available.
15. For a detailed review of the various large international research projects and programs currently under way, see Organisation for Economic Co-Operation and Development, OECD Megascience Forum (1993), *Megascience and Its Background*, Paris. See also the OECD Megascience Forum Web site at <http://www.oecd.org/dsti/mega/>.
16. 15 United States Code, Section 5652 (1992).
17. See General Accounting Office (1990), *Environmental Data—Major Effort Is Needed to Improve NOAA's Data Management and Archiving*, Washington, D.C.; and General Accounting Office (1990), *Space Operations—NASA Is Not Archiving All Potentially Valuable Data*, Washington, D.C. It should be noted that both agencies have taken significant measures to rectify these past problems.
18. National Research Council (1995), *Preserving Scientific Data*, note 3.
19. National Research Council (1995), *Preserving Scientific Data*, note 3, at pp. 47-48.
20. See Gary Taubes, (1996), "Science Journals Go Wired," *Science* 271(February 9):764; and UNESCO Expert Conference on Electronic Publishing in Science (1996), ICSU Press at <http://www.lmcp.jussieu.fr/~fabrice/icsu/information/index.html>. See also, Steve Hitchcock, Leslie Carr, and Wendy Hall, "A Survey of STM On-line Journals 1990-95: The Calm Before the Storm," at <http://journals.ecs.soton.ac.uk/survey/survey.html>.
21. <http://www.aip.org:80/>.
22. <http://www.iop.org/>.
23. See <http://eij.gsfc.nasa.gov>.
24. See, for example, the NASA-funded Astrophysics Data System Abstract Service at <http://adswww.harvard.edu/ads_abstracts.html>.
25. See Richard T. Kouzes, James D. Myers, and William A. Wulf (1996), "Collaboratories: Doing Science on the Internet," *IEEE Computer* 29(8):40-46.
26. For additional insights in this area, see the *Proceedings of the '96 UNESCO Conference on Electronic Publishing in Science*, held at the UNESCO Headquarters in Paris, February 19-23, 1996. A summary of the results from that conference was presented by D.F. Shaw (1996), "Electronic Publishing in Science," *Science International*, ICSU Paris, May, pp. 1-3.
27. See Nahum Gershon and Judith R. Brown (1996), "Computer Graphics and Visualization in the Global Information Infrastructure," a Special Report in *IEEE Computer Graphics and Applications*, March, pp. 60-75; and Robert Braham (1995), "Math & Visualization: New Tools, New Frontiers," a Focus Report in *IEEE Spectrum*, November, pp. 19-65.
28. Canadian Global Change Program (1996), "Data Policy and Barriers to Data Access in Canada: Issues for Global Change Research," The Royal Society of Canada, Ottawa. National Research Council (1993), *1992 Review of the World Data Center A for Rockets and Satellites*, National Space Science Data Center, Board on Earth Sciences and Resources, National Academy Press, Washington, D.C.; National Research Council (1992), *Toward a Coordinated Spatial Data Infrastructure for the Nation*, Board on Earth Sciences and Resources, National Academy Press, Washington, D.C.; National Academy of Public Administration (1991), *The Archives of the Future: Archival Strategies for the Treatment of Electronic Databases*, A report for the National Archives and Records Administration; General Accounting Office (1990), *Environmental Data—Major Effort Is Needed to Improve NOAA's Data Management and Archiving*,

Washington, D.C.; General Accounting Office (1990), *Space Operations—NASA Is Not Archiving All Potentially Valuable Data,* Washington, D.C.; National Research Council (1990), *Spatial Data Needs: The Future of the National Mapping Program,* Board on Earth Sciences and Resources, National Academy Press, Washington, D.C.; National Research Council (1988), *Geophysical Data: Policy Issues,* Committee on Geophysical Data,, National Academy Press, Washington, D.C.; National Research Council (1988), *Selected Issues in Space Science Data Management and Computation,* Space Science Board, National Academy Press, Washington, D.C.; National Research Council (1986), *Atmospheric Climate Data: Problems and Promises,* Board on Atmospheric Sciences and Climate, National Academy Press, Washington, D.C.; National Research Council (1986), *Issues and Recommendations Associated with Distributed Computation and Data Management Systems for the Space Sciences,* Space Science Board, National Academy Press, Washington, D.C.; J.K. Haas, H.W. Samuels, and B.T. Simmons (1985), *Appraising the Records of Modern Science and Technology: A Guide,* Massachusetts Institute of Technology, Cambridge, Mass.; National Research Council (1984), *Solar-Terrestrial Data Access, Distribution and Archiving,* Space Science Board and Board on Atmospheric Sciences and Climate, National Academy Press, Washington, D.C.; National Research Council (1982), *Selected Issues in Space Science Data Management and Computation,* Space Science Board, National Academy Press, Washington, D.C.
29. Committee on the Future of Long-term Ecological Data (FLED), (1995), *Final Report of the Ecological Society of America,* Katherine L. Gross, Chair, Ecological Society of America, Washington, D.C.; National Research Council (1993), *A Biological Survey for the Nation,* Committee on the Formation of the National Biological Survey, National Academy Press, Washington, D.C.
30. National Research Council (1986), *Toward a Geosphere-Biosphere Program,* National Academy Press, Washington, D.C.; National Research Council (1988), *Toward an Understanding of Global Change: Initial Priorities for U.S. Contributions to the International Geosphere-Biosphere Program,* National Academy Press, Washington, D.C.; and National Research Council (1990), *Research Strategies for the U.S. Global Change Research Program,* National Academy Press, Washington, D.C.
31. Additional information on the Global Climate Observing System can be found at the World Meteorological Organizations's Web site at <http://www.wmo.ch/web/gcos/gcoshome.html>; and see <http://www.wmo.ch/web/www/www.html> for the World Weather Watch, and <http://www.wmo.ch/web/arep/gaw.html> for the Global Weather Watch. See <http://www.unesco.org/ioc/goos/iocgoos.html> for the Global Ocean Observing System and <http://www.wsl.ch/wsidb/gtos/gtos.html> for the Global Terrestrial Observing System.
32. See the Carbon Dioxide Information Analysis Center's Web site at <http://cdiac.esd.ornl.gov/cdiac>.
33. For extensive discussion of data quality control and assurance procedures and recommendations in the context of interdisciplinary environmental research, see National Research Council (1995), *Finding the Forest in the Trees: The Challenge of Combining Diverse Environmental Data,* National Academy Press, Washington, D.C.
34. For a general overview of issues and requirements in archiving digital data and information, see Task Force on Archiving of Digital Information (1995), *Preserving Digital Information,* the Commission on Preservation and Access and the Research Libraries Group, Inc., at http://lyra.rlg.org./Arch TF/>.
35. See, for example, Committee on the Future of Long-term Ecological Data (FLED), (1995), *Final Report of the Ecological Society of America,* Katherine L. Gross, Chair, Ecological Society of America, Washington, D.C.
36. National Research Council (1995), *Preserving Scientific Data,* note 3.
37. National Research Council (1994), *Facing the Challenge: The U.S. National Report to the IDNDR World Conference on Natural Disaster Reduction,* Yokohama, Japan, May 23-27, 1994,

U.S. National Committee for the Decade for Natural Disaster Reduction, National Academy Press, Washington, D.C.; National Research Council (1991), *A Safer Future: Reducing the Impacts of Natural Disasters,* U.S. National Committee for the Decade for Natural Disaster Reduction, National Academy Press, Washington, D.C.

38. National Research Council (1995), *Finding the Forest in the Trees,* note 33.
39. See, generally, National Research Council (1992), *Toward a Coordinated Spatial Data Infrastructure for the Nation,* Board on Earth Sciences and Resources, National Academy Press, Washington, D.C.
40. Secondary users, such as researchers in other fields, policymakers, educators, and the general public, do not collect and create data sets, but they perform tasks with, analyze, and interpret the data. For a discussion of distinctions between user categories, see National Research Council (1995), *Study on the Long-term Retention of Selected Scientific and Technical Records of the Federal Government—Working Papers,* National Academy Press, Washington, D.C.
41. See National Research Council (1995), *Finding the Forest in the Trees,* note 33, and *Preserving Scientific Data,* note 3. For general information on metadata issues, see the Lawrence Livermore National Laboratory Metadata and Data Management information page at <http://www.llnl.gov/liv_comp/metadata/metadata.html>.
42. For more information regarding IGBP-DIS data activities, see <http://www.cnrm.meteo.fr:8000/igbp/outline.html>. Additional information is provided in the "Summary Report on the 7th IGBP-DIS Scientific Steering Committee Meeting Manual" (1996) at <http://www.cnrm.meteo.fr:8000/igbp/meetings summary_rep_ssc_feb96_ver_html.html>. See also the NASA Global Change Master Directory for another example of a successful on-line indexing effort at <http://gcmd.gsfc.nasa.gov/cgi-Bin/pointers/>.
43. Michael Carlowicz (1997), "New Data from Cold War Treasure Trove," *EOS,* American Geophysical Union, Vol. 78, no. 9, March 4, p. 93.
44. See "Corona: America's First Satellite Program" (1995), Kevin C. Ruffner, ed., Central Intelligence Agency History Staff Center for the Study of Intelligence, Washington, D.C.; and Robert A. McDonald (1995), "Opening the Cold War Sky to the Public: Declassifying Satellite Reconnaissance Imagery," *Photogrammetric Engineering and Remote Sensing,* pp. 385-390. For a listing of declassified satellite data products, including information about missions, dates, and resolution, see the United States Geological Survey's EROS Data Center Web site at <http://edcwww.cr.usgs.gov/glis/hyper/guide/disp>.
45. See William J. Broad (1996), "Anti-Sub Seabed Grid Thrown Open to Eavesdropping," *New York Times,* July 2, p. C1.
46. See National Research Council (1988), *Ozone Depletion, Greenhouse Gases and Climate Change: Proceedings of a Joint Symposium by the Board on Atmospheric Sciences and Climate and the Committee on Global Change,* National Academy Press, Washington, D.C.
47. National Research Council (1995), *On the Full and Open Exchange of Scientific Data,* National Academy Press, Washington, D.C. The appendix to the report also presents a collection of other similar supporting policy statements.
48. See Information Infrastructure Task Force (1994), *The Global Information Infrastructure: Agenda for Cooperation,* Washington, D.C., at <http://www.iitf.nist.gov/documents/docs/gii/giiagend.html>.
49. J.R.G. Townshend, C. Justice, W. Li, C. Gurney, and J. McManus (1991), "Global Land Cover Classification by Remote Sensing: Present Capabilities and Future Possibilities," *Remote Sensing and the Environment* 35:243-355.
50. See T.R. Loveland, J.W. Merchant, D.O. Ohlen, and J.F. Brown (1991), "Development of a Land Cover Characteristics Database for the Conterminous U.S.," *Photogrammetric Engineering and Remote Sensing* 57:1453-1463.
51. These points have been discussed in some detail for the field of materials databases, where the high volume and critical importance of metadata, the broad scope of the materials field, the rich

SCIENTIFIC ISSUES 107

vocabulary of materials technology, and the international character of materials information give special importance to the subjects. See J.H. Westbrook and W. Grattidge (1992), "Terminological Standards for Materials Databases," in *Computerization and Networking of Materials Databases*, Vol. 3, T.J. Barry and K.W. Reynard, eds., American Society for Testing and Materials, Philadelphia, pp. 15-33.

52. D.L. Hawksworth, B.C. Sutton, and G.C. Ainsworth (1983), *Dictionary of the Fungi* (including the Lichens), seventh edition, Commonwealth Mycological Institute, Kew, Surrey, England.

53. College of American Pathologists, Committee on Nomenclature and Classification of Disease (1965), *Systematized Nomenclature of Pathology*, first edition, American Cancer Society and American Medical Association, Chicago.

54. A reference with numerous examples from the field of chemistry is J.H. Westbrook (1993), "Problems in the Computerization of Chemical Information: Capture of Tabular and Graphical Data," *Journal of Chemical Information and Computer Sciences* 33:6-17.

55. See <http://www.nist.gov/srd/>.

56. See the recommendations in National Research Council (1995), *Preserving Scientific Data*, note 3.

57. For example, the American Association for the Advancement of Science sponsors the Project for African Research Libraries in partnership with U.S. scientific societies to provide subscriptions for core scientific and technical journals in 35 sub-Saharan African countries (see <http://www.aaas.org/international/ssa-1.htm> for general information on international programs). For UNESCO's programs on the advancement, transfer, and sharing of knowledge in the natural sciences, see also <http://www.unesco.org/ch-intern/programmes/science/highlights.html>.

58. These statistics vary over time and according to country and discipline, and are available for only a few major countries. See *Science & Engineering Indicators* (1996), National Science Board, Washington, D.C., pp. 2-28 to 2-30. The statistics indicate that 35 to 75 percent of foreign graduate students surveyed intend to stay in the United States upon completion of their studies.

59. For an overview of potential educational activities to improve the management of scientific and engineering data, see National Research Council (1986), *Improving the Treatment of Scientific and Engineering Data Through Education*, National Academy Press, Washington, D.C.

60. This is not an exhaustive list of organizations that provide assistance to scientists in developing countries. Several organizations, such as the International Development Research Centre Library, provide extensive links to Internet sites related to international development (see <http://www.irdc.ca/library/world/world.html>).

61. USAID focuses on regional activities, such as the African Data Dissemination Service (ADDS), which is conducted in conjunction with several private organizations, the Office of Arid Land Studies at the University of Arizona, NASA, and NOAA. An example of an ADDS project is the Famine Early Warning System, which, with the help of the USGS EROS Data Center, provides information about potential famine situations to allow for proactive initiatives to prevent famine (see <http://edcsnw4.cr.usgs.gov/adds/general/> for additional information on ADDS). Other USAID activities, such as AfricaLink and the Leland Initiative, provide network connections and information management to Africa (see <http://www.info.usaid.gov/regions/afr/> for a description of these and other regional programs in Africa). The USAID-sponsored U.S.-Russian NGO Cooperation Project provides small grants and equipment for individuals and institutions to link to an environmental e-mail network in Central Asia and the West Newly Independent States (see <gopher://gaia.info.usaid.gov:70/00...enis_reg/nis.factsheet/enviro3.txt>).

62. For example, the NASA Pathfinder project uses Landsat images to determine forest land cover and change for three quarters of the world. The USDA Foreign Agriculture Service provides support to the Consultative Group on International Agriculture Research (CGIAR) through the prediction of global production of major grains. For additional information on both programs, see the *Proceedings of a Workshop on the Use of Remote Sensing Technologies and GIS*

Database in CGIAR Centers (1995), Environment and National Resources Information Center (see <http://www.info.usaid.gov/environment/enric/special/cgiar.htm>).
63. See <http://www.state.gov/www/global/oes/envir.html>.
64. See <http://www.unsystem.org> for the official listing of the United Nations system of organizations' Internet servers. Numerous initiatives within the U.N. programs and specialized agencies directly assist scientists in developing countries through a variety of mechanisms. One example is the Sustainable Development Network Programme of the UNDP, which links government organizations, the private sector, universities, NGOs, and individuals in developing countries through electronic networks for the purpose of exchanging information on sustainable development (see <http://www3.undp.org>). Refer to the programs' and agencies' home pages for further details on other U.N. initiatives.
65. For example, the OAS RedHUCyT program is a hemisphere-wide interuniversity scientific and technological information network created in 1991 with the objective to connect OAS member countries to the Internet, "integrating an electronic network for the exchange of scientific and technological information among professors, researchers, and specialists, at different universities in the member states" (for additional information, see <http://www.oas.org/EN/PROG/RED/covere.htm>). OAS also sponsors a regional scientific and technological development program, which carries out a number of multinational and national projects that provide member states "with an opportunity to share experiences, to provide . . . mutual support and to engage in joint activities to further the advancement of science and technology and to promote integral development" (see <http://www.oas.org/EN/PROG/pa26e.htm>).
66. See the Institute for Baltic Studies' Web site at <http://www.ibs.ee/dollar/fw4/wp/dev.html>.
67. <http://gds.esrin.esa.it:80/559DE416/T0xc1cce622_0x00029290>.
68. The Third World Academy of Sciences (TWAS) was founded in 1983 to support scientific research in developing countries through provision of research grants, spare parts for scientific equipment, books and journals, and fellowships. See <http://www.ictp.trieste.it/TWAS.html/> for a description of TWAS activities and programs. The organization not only is closely coupled with the U.N., but also collaborates with the International Council of Scientific Unions (see next note).
69. The International Council of Scientific Unions (ICSU) was founded in 1933 to "bring together natural scientists in international scientific endeavor." ICSU works closely with UNESCO, WMO, FAO, and UNEP through formal or ad hoc collaborations (see <http://www.lmcp.jussieu.fr/~fabrice/icsu/> for additional information on ICSU). ICSU's Committee on Science and Technology in Developing Countries (COSTED) was created in 1966 to stimulate international scientific and technological cooperation in developing countries. It is a joint initiative co-sponsored by UNESCO and was merged with the International Biosciences Network, an activity with similar objectives, in 1994. For additional information on COSTED and its activities, see G. Thyagara (1995), "Cooperative Research for Development Is COSTED's Aim," *The Hindu On-line* (<http://www.webpage.com/hindu/960113/22/0820a.html>). ICSU also works to assist scientists in developing countries through its scientific unions and interdisciplinary committees; for example, CODATA recently established the Task Group on Outreach, Education, and Communication, which promotes collaboration, scientific information exchange, and technology transfer for individual scientists and technologists in developing nations.
70. Founded in 1972, the International Foundation for Science (IFS) provides support (in the form of research grants, equipment, regional workshops and training courses, and travel grants) to young scientists in developing countries in the following research areas: aquatic resources, animal production, crop science, forestry/agroforestry, food science, and natural products. See <http://ifs.plants.ox.ac.uk/ifs/> for additional information about IFS activities and programs.
71. See <http://www/ciesin.org> for additional information about CIESIN's programs and services.

72. See <http://www.std.com/sabre/SAP/sap.info.html> for additional information on the Sabre Foundation.
73. See <http://www.isf.ru/index-isf.html> for additional information about the International Science Foundation and its various programs that assist scientists, such as its Library Assistance Program and the Telecommunications Program.
74. See <http://info.irex.org> for additional information about IREX programs.
75. See <http://vita.org>.
76. See American Association for the Advancement of Science, *Science and Technology in the Americas: Perspectives on Pan American Collaboration* (1994), 2nd ed., E. Jeffrey Stann, ed., AAAS, Washington, D.C.
77. See James M. Musser (1996), "Molecular Population Genetic Analysis of Emerged Bacterial Pathogens: Selected Insights," *EID*, 2(1), January-March.
78. See *Biodiversity Prospecting: Using Genetic Resources for Sustainable Development* (1993), Reid et al., World Resources Institute, Washington, D.C., 350 pp.; and "Bioprospecting/Biopiracy and Indigenous Peoples" (1994) *RAFI Communique*, Nov./Dec.
79. Personal communication from a member of the staff of the Brazilian Embassy, Washington, D.C., 1995.
80. See the recommendations in National Research Council (1995), *Preserving Scientific Data*, note 3.

4

Data from Publicly Funded Research—The Economic Perspective

The most striking theme throughout this report is how progress in information technologies has changed the way science is accomplished. It has enabled the generation, processing, storage, and distribution of quantities of data undreamed of even a decade ago.

Sensing systems (e.g., Earth observation satellites, the Hubble Space Telescope, ground-based radars) and other forms of automated data generation (e.g., genome studies) produce enormous amounts of useful data, enabling scientists to study natural phenomena at a much greater level of detail and granularity than was hitherto possible. Science and scientists have been the main drivers of this highly sophisticated and often very expensive technology, using it to push forward the frontiers of knowledge in their respective disciplines. The continuous increases in the processing power available to analyze the data are as crucial to this evolution as the improvements in data generation capabilities. Surprisingly, these increases have not come principally from more powerful supercomputers, but rather from cheaper, more powerful workstations and PCs available throughout the scientific community. The development of inexpensive mass storage media has ensured that the preservation of these vast quantities of data, both processed and unprocessed, is both possible and affordable. Finally, the most profound change in technology has been the worldwide growth of the Internet, with its potential to make data from anywhere in the world available anywhere else in the world, instantaneously, and, increasingly, in large quantities.

These four factors, taken together, have revolutionized the way science is conducted, making it truly global. Perhaps most interesting is that this progress has changed the way scientists communicate with each other. Physical scientists are leading the way among scholars in the publication of electronic journals, compressing the time between discovery and communication of the results. This phenomenon is accelerating the already rapid pace of discovery and innovation, as the cycle time of discovery, communication, and next discovery is reduced. The committee uses the term "digitization of science" as a shorthand for this phenomenon.

As a consequence, the flow of scientific data and information has been improved, as the cost of publication and of access to information has been drastically reduced. While not all scientists in every country have full access to modern PCs and fast Internet connections, these technologies are becoming widespread and are likely to be ubiquitous in the near future.

This digitization of science has occurred contemporaneously (and coincidentally) with the demise of the great power rivalry of the Cold War. Russian and U.S. scientific relations have become less heavily dominated by security considerations, and this factor also has led to an increase in the availability and transfer of scientific data, as noted in Chapter 3.

At the same time, there have been fundamental changes in how governments in many countries see their role relative to markets. Budget pressures, plus the evident success of market economies, have led many governments to privatize activities previously delivered via the public sector, in hopes of relieving the burden on taxpayers while improving the allocation of economic resources. These pressures have begun to be felt in the area of scientific data; for example, in the United States, Landsat remote sensing was privatized in the mid-1980s, and some European countries have strongly urged limits on the sharing of meteorological and other data in order to protect the data markets for their government monopolies.

THE TREND TOWARD MARKETS—GOOD OR BAD FOR SCIENCE?

To researchers and educators in the natural sciences, this pressure toward privatization and commercialization of scientific data is of great concern. Many fear that scientific data, the lifeblood of science, will be priced beyond their means, especially in less developed countries. It is argued, correctly, that the conduct of scientific research, including the maintenance and distribution of scientific data, is a public good, provided for by government funding (see Box 4.1). This traditional model[1] has worked well in the past, and many scientists[2] are of the view that privatizing the distribution of scientific data will impede scientific research.

To the economist, this concern at first seems misplaced. While the conduct of scientific research certainly is a public good, one might consider the maintenance and distribution of scientific data as the provision of one of the commodities used by scientists. This view makes scientific data analogous to the chemicals, computers, and travel that each scientist is free to buy or not, as they best

BOX 4.1
What Is a Public Good?

Economic analysis recognizes that not all goods can be easily transacted through markets. There is an important class of economic goods, called *public goods*, in which markets may not work well. The term "public good" does not refer to something that is "good for the public." It refers to a product or service possessed of certain properties that lead to collective consumption or production, rather than private consumption or production. A public good is characterized by two attributes, *nondepletability* and *nonexcludability* (nonappropriability). Nondepletability means that the product in question cannot be used up and is thus available to additional persons. If I eat a slice of cheese, less remains available for others to consume. However, if I use the latest statistics on the number of employees in the steel industry or the number of persons infected by some disease, those statistics remain just as available as before for use by others—their supply is not depleted by additional use or additional users. Nondepletability is the main reason economists conclude that free use of public goods can be justified—there is no social cost when another person uses them, and there is no justification for the disincentive to their use that is constituted by a substantial price for that use. This, then, is the user side of the dilemma of provision of public goods—economic efficiency calls for a zero or very low price for their use, but private enterprise cannot be expected to provide a costly and valuable service without charge.

The second attribute, nonexcludability, is the supplier side of the problem. Nonexcludability means that the good in question produces benefits from which others cannot be excluded and which cannot easily be constrained only to those who pay. A classic case is that of national defense; defending one American involves defending all Americans. One cannot supply the service to some people but exclude others. Information also possesses this attribute, although somewhat less fully than national defense. Once information is provided to some, it is likely, but not certain, to leak out to others. Absence of excludability makes it very difficult for the provider of such a service to collect reimbursement for the cost of supplying the service. That is why, economists conclude, it is often difficult for any supplier other than government to provide certain types of public goods.

It is generally accepted that scientific research itself has strong public-good attributes, in that the knowledge produced by such research is freely available to all (i.e., nonexcludable) and provides social and economic benefits to members of society far beyond those who produce it and those who pay for it. Such goods are usually provided by governments—or they are not provided at all.

The public-good nature of science is not limited to any particular nation. The scientific endeavor has traditionally been and will continue to be a global enterprise; the success of this endeavor depends critically on the global community of scientists, and their ability to work with innovators, implementers, and users. To the extent that this global interchange is restricted or inhibited, the long-run contribution of science to the U.S. economy will decrease. Thus, the United States has an interest not only in a healthy domestic scientific community, but also in a robust global scientific community.

However, the issue in this report is not scientific research, but rather the data that science generates, either as input to scientific research (e.g., data from meteorological satellites) or as output from scientific research (e.g., description of a

> gene sequence). Clearly, scientific data have some aspects of a public good. On the other hand, scholarly journals have been copyrighting scientific articles for years (thereby privatizing them) without impeding the flow of science. Certainly, the provision of scientific data has important spillovers: future researchers within the field benefit from these data, researchers in other fields also may benefit (e.g., medical researchers benefit from the provision of biological research data), and commercial firms (e.g., pharmaceutical firms) may benefit as well. Unfortunately, uncertainty about who the ultimate beneficiaries are, which appears to be fundamental to science, precludes asking those beneficiaries of spillovers to pay.

serve the scientist's needs. In this traditional market model, both consumers and researchers would be better served by suppliers anxious to survive by supplying the most desired data set traits: reliability, accuracy, timeliness, and so forth. Instead of the government providing financial support to data centers, why not give the funds to individual researchers, who can then make choices among data suppliers who best serve their needs? To many economists, therefore, privatization appears at first blush to be a positive development for science.

A more careful economic analysis of the maintenance and distribution of scientific data, however, suggests that a somewhat different market model is more appropriate here, for a number of reasons. In some cases, the (public good) scientific research is tightly tied to the collection, maintenance, and distribution of the data generated from the research. For example, the Hubble Space Telescope (HST) project (and other space science observatories), clearly a public good, collects, maintains, and distributes the data from the HST as part and parcel of the project itself. In this case, the basic research is vertically integrated with data distribution, and separating the two functions would create more difficulties than keeping them integrated.

In other cases, frequently overlapping with the conditions described above for the HST project, the contributors of scientific data are the same as the consumers of the data, all of whom are members of the same relatively small research community. This is especially so in relatively esoteric areas of fundamental research, such as high-energy physics or paleontology. This model is closer to that of a family or clan, in which exchange is not monetized but depends on social norms specifying expected and well-understood levels of contribution. Imposing a market price system in such a situation could not only be countercultural, but also counterproductive. For example, administrative expenses of instituting such a system might well be higher than the revenues realized. In yet other cases, the market for scientific data is not large enough to support more than a single supplier, which would mean that the data either would not be provided by the market or would be provided under monopoly conditions.

An additional and more subtle difficulty arises from the nature of the funding of scientific research. There is no question that the public-good nature of fundamental scientific research requires public funding,[3] and this includes ensuring that researchers have the necessary inputs such as scientific data.[4] Privatizing data distribution would not change the requirement that data acquisition be publicly funded; it would simply change the locus of funding from the supplier/distributor of data (i.e., publicly funded and operated data centers providing data either free or at low cost to researchers) to the consumer of data (i.e., the researchers would have a "data budget" as part of their grant, which they would use to shop among suppliers for their data needs).

But are these two modes of funding equivalent? The second mode might be thought by economists to be superior, in that it puts the financial power in the hands of the consumer. However, the fact that both involve public funding, subject to year-to-year political vagaries, suggests a different view. In the first mode, funding is directed to government agencies or large research institutions, which thereby have an interest in continued funding and can make the case to their legislators for such funding. In the second mode, funding is directed to individual researchers, who, while they have an interest in ensuring continued funding for purchasing data, clearly do not have the political ability to protect this funding from political pressures to reduce or eliminate it. This inability is a major factor underlying scientists' concerns that if the distribution of scientific data were privatized, the increases in research grants to enable their data purchases would soon disappear, leaving researchers and perhaps their universities to pick up the bill. The economic problem here is that *government cannot commit to future funding of researchers for buying data.* Neither Congress nor the Administration can make credible commitments for future funding. Ensuring that large institutions have the means and the access to argue the political case for scientific data will increase the likelihood that future funding of such data will be made available.[5] The committee believes that direct appropriation or block grant support to institutions with broad responsibilities for data management, retention, and distribution, while not assured, is typically more stable and secure, and fortified by an institutional memory to recognize and support the continued utility of archived data. Thus, a strong case can be made for funding (subsidizing) institutions to be data distributors as part of the infrastructure for government research in cases where both (1) the long-term availability of the data is essential for the conduct of research, and (2) there is no guarantee of continued financial support of the user community for acquiring data.

DETERMINANTS OF THE STRUCTURE OF SCIENTIFIC DATA DISTRIBUTION

How best to distribute scientific data depends on several economic properties of the underlying science and scientific community that both generates and uses the data, as well as other uses for the data. To determine these properties, the key questions are as follows:

- *Does the scientific research depend on a substantial public investment in one or more facilities that generate the data of interest?* The Hubble Space Telescope is a clear example of a single costly facility, the sole purpose of which is to generate basic scientific data for its useful life. Another example is weather or other earth observing satellites, which constitute major facilities in their own right, but that also contribute to a common broader research data set. In these cases the data have significant nonresearch applications as well, which are a mix of public good and commercial uses. The collection, processing, and distribution of data typically are most efficiently vertically integrated in the same program. The costly observational facility is usually provided by government (e.g., NOAA, NASA, ESA), and so the distribution of the resulting data (at least the minimally processed data) is best handled by the same agency.
- *Is the (non-facilities-based) scientific research coordinated across researchers, possibly in different countries?* An example of this situation is the Human Genome Project. Individual researchers throughout the world contribute to this effort; the results of each individual research project are made available to all researchers. In fact, the maintenance of common data sets available to all is what defines this as a project. In such cases, there typically is a lead government agency with responsibility for the entire project that also takes responsibility, either directly or indirectly, for providing for data collection and distribution. Though not facilities-based, a common repository of information and scientific data is essential to the conduct and usefulness of the entire project. An important element of such a repository is that it is the mechanism by which researchers communicate with each other. In some cases, such as paleoclimatological research, contributors and users are largely the same, and the repository acts solely as the means of professional exchange. In this situation, ensuring that the data are freely available is part of the project itself. The responsible agency may provide the distribution facility, or it may fund a university, consortium of universities, or other nongovernmental organizations (NGOs) to operate the facility.
- *Is the community of users roughly the same as the community of contributors?* As with the paleoclimatology or astronomy examples above, the distribution of scientific data is best viewed as a sharing of results within a community, rather than as a market opportunity. If the community of users is much broader than those making contributions, then distributing scientific data is also a publication function, possibly to private concerns. A good example of this situation is the Human Genome Project, for which the repository of research results is of interest to others beyond the immediate research community, such as pharmaceutical companies.

In cases where the user community is much larger than the contributor community, governments or NGOs may still wish to make the information available to all at no cost. This was the case for the World Meteorological Organization's World Weather Watch, discussed in the previous chapter. However, for budget reasons nonscientific users may be required to pay for the data. Under this arrangement, some form of price discrimination or product differentiation may be required, in which scientific researchers can acquire the data free or at very low cost, while nonresearch users are charged for the data.[6] This can be done by the responsible agency itself. However, the agency first must determine whether the transaction costs of establishing and maintaining discriminatory prices exceed any extra income so derived, particularly for data from narrowly focused basic research projects. Obviously, an agency should implement a price discrimination scheme only if the efficiency gains (not just the revenue gains) outweigh the transaction and administrative costs of doing so. Price discrimination, in practice, will be worth the effort only with a sufficiently large commercial user base.

- *Is the user community large enough to support more than one data distributor?* In many cases, a particular scientific data set is likely to be of interest to only a few scientists and practitioners, and a private[7] market may support only one distributor, due to scale economies.[8] In such cases, privatizing data distribution will result in a private monopoly with no incentive to support the public interest, replacing a public monopoly that does have such a commitment. Of course, government monopolies that do not sustain activities providing public goods are no better than private monopolies.

As a quick reference, Table 4.1 lists the relevant properties mentioned above for each of a number of diverse scientific data sets.

PRIVATIZATION: WHEN DOES IT MAKE SENSE?

Given the unique nature of scientific data, it might appear that government (or NGO subcontractor) distribution of such data is always the correct choice. However, there may be opportunities for private firms to reformat, enhance, and market these publicly available data in new, added-value forms. Private firms may have capabilities not available to government agencies or NGOs that would add value for various end users. For example, NOAA distributes weather data to all users on equal terms, including to commercial firms, some of which package this information and provide forecasting services to the public via mass media. While NOAA clearly has a huge advantage in technology and meteorological science, the commercial firms have an equally clear advantage in packaging the information for maximum public impact.[9] Thus, any private supplier who requests access to scientific data should be given it and should be permitted to go into competition with other suppliers *and with the government itself.* The marketplace will then determine the best package of service, support, and reliable data for users, including scientific researchers.[10] Fortunately, there are many cases in addition to that for weather data in which private sector distribution of value-

TABLE 4.1 Properties Relevant to Distribution for a Number of Scientific Data Sets

	Facilities-based	Single/Multiple Activities	No. of Users Equal to (=) or Greater Than (>) Contributors?	Observational/Experimental	Raw/Processed	National/Global	Major Commercial Interest?	Pricing Basis[a] Discrimination?
Genbank—NIH/NLM/NCBI	N	M	>	E	P (submittal/retrieval software)	G	Yes	Internet access free; CD-ROM @MC
Genome Database—Johns Hopkins University, funded by DOE, NIH, Japan S&T	N	M	>	E	P	G	Yes	Free
Hubble Space Telescope—NASA/ESA (operated by AURA)	Y	S	= (principally space researchers)	O	P	G	No; occasional use by publishers	Free
National Space Science Data Center—NASA	Y	M	=	O	R/P	G	No	@MC
European Space Information System	Y	M	=	O	P (browser/imaging software)	G	No	Free
National Center for Atmospheric Research—Not-for-profit, NSF funding	Y	M	=	O	R/P	N	Little	@MC; additional services available

(*continues*)

TABLE 4.1 Continued

	Facilities-based	Single/Multiple Activities	No. of Users Equal to (=) or Greater Than (>) Contributors?	Observational/Experimental	Raw/Processed	National/Global	Major Commercial Interest?	Pricing Basis[a] Discrimination?
T-2 Nuclear Information Service—UC,[b] DOE	Y	M	=	E	P	G	No	Free
Australian Oceanographic Data Centre—Australian Dept. of Defense	Y	M	>, but mostly researchers	O	P (free software)	G	Some	Free as "public good," @MC for commercial
ESA—Information Retrieval Service—European Union Space Agency	Y	M	>	O and E; includes financial data	P	G; focused on European Union	Perhaps some	Appears to be > IC
Earth Resources Observation Systems—USGS (DOI)	Y	M	>	O	R/P	G	Some; used by government agencies	Varies; ranges from free to > IC
Scripps Institution of Oceanography—Not-for-profit	Y	M	=	O	P	G	Little	@MC

University of Alaska GeoData Center—University/State funds	Y	M	>; mostly public	O	R	N	Little	@MC
Paleoclimatology World Data Center A—NGDC NOAA/DOC	N	M	=	O	R/P	G	No	Free for direct contributors; @MC otherwise
National Climatic Data Center—NOAA/DOC	Y	M	>	O	P	G	No	@MC; careful accounting for analysis to determine MC
Incorporated Research Institutions for Seismology (IRIS)—University consortium; NSF funding	Y	M	>	O	P	G	No	Free
Properties of Intermetallic Alloys—Purdue University CINDAS/MIAC (DOD)	N	M	>	E	P	G	Yes	Printed books fairly expensive; higher cost for foreign users
Evaluated Nuclear Data File/B (DOE)	N	M	>	E	P	G	No	Free

[a] Pricing basis: IC = incremental cost; MC = marginal cost, or cost of reproduction.
[b] Abbreviations and acronyms are defined in Appendix A.

> **BOX 4.2**
> **FAME: A Private-Public Sector Success**
>
> An example of how a market can be made from subsidized data generation is the Fatty Acid Methyl Ester, or FAME, system of identifying bacteria. The fatty acids are extracted from the bacteria to be identified, made volatile by methyl esterification, and subjected to chromatography. The resultant chromatogram yields a profile pattern that is analyzed statistically to identify the bacteria. The profiles and the statistics make up the database of interest. The original work was done, and the database was compiled, at the Centers for Disease Control and Prevention (CDC) in Atlanta, Georgia. The CDC staff is augmenting and actively using the database in fulfilling its mission. Much of the database is included in a publication for use by bacteriologists, especially in clinical microbiology laboratories.[1]
> A commercial company, MIDI,[2] adapted the CDC system, developed its own proprietary database, and apparently has been successful in commercializing the system. The two databases have gone their separate ways, competing in the intellectual, but not the economic, marketplace.
>
> ---
> [1]W.S. Weyant, C.W. Moss, R.E. Weaver, D.G. Hollis, J.G. Jordan, E.C. Cook, and M.I. Daneshvar (1996), *Identification of Unusual Pathogenic Gram-negative Aerobic and Facultatively Anaerobic Bacteria*, Williams and Wilkins, Baltimore, Md.
> [2]Located at 115 Barsdale Professional Center, Newark, DE 19711.
>
> SOURCE: Micah Krichevsky, Bionomics International, Bethesda, Md.

added data has worked to the benefit of science and the broader public. The Fatty Acid Methyl Ester (FAME) database in microbiology (Box 4.2) is a case in which a private firm offers alternatives to the data available from government. The committee lists below some necessary conditions for the complete privatization of scientific data distribution to be an appropriate option:

- *Can the distribution of data be separated easily from their generation?* For the HST, the answer is "no"; for the Human Genome Project, the answer is likely "yes."
- *Is the scientific data set used by others beyond the research community.* Again, for the HST, the answer is "no"; for the Human Genome Project, the answer is likely "yes."
- *Is the potential market large enough to support several data distributors* If so, then the resulting private market could be competitive, and privatization could be helpful to scientists and others. If not, then privatization could lead to monopolization, which would likely be detrimental to the interests of science.
- *Is it easy to discriminate prices or differentiate products between scientific users and other users? If this is possible, can low prices be mandated contractually for government-funded data for scientific users?* If so, then it is likely that scientists will obtain the needed data on more favorable terms than their colleagues in private industry with more resources.

- *Is it costly to separate the distribution of data to scientists from their distribution to other users, such as commercial users?* For small or esoteric research communities, economies of scale in data distribution may make this separation costly.

If all these questions can be answered "yes," then privatizing the distribution of scientific data should be an option to be considered. Privatizing data distribution might appear to be attractive to a budget-constrained government agency. It certainly removes the cost of maintaining and distributing the data and may even bring in revenues that can help the agency fulfill other aspects of its mission.

BOX 4.3
The Failed Privatization of Landsat

The story of what happened to the Landsat system when it was privatized is instructive. Landsat is a series of remote sensing satellites, the first of which was launched by NASA in 1972. The Carter Administration proposed the privatization of Landsat in 1978. This led in the early 1980s to a transfer of responsibility for the system to NOAA, which attempted to build a customer base for Landsat's data products. Under both NASA and NOAA management, Landsat images were made available to all users at the marginal cost of reproduction.

The privatization process was accelerated during the Reagan Administration. Congress passed the Land Remote-Sensing Commercialization Act in 1984, setting forth the general provisions for privatization of the system. In 1985, the Earth Observation Satellite (EOSAT) Co. (a joint venture of Hughes and RCA) was awarded the contract, along with an additional $250 million and a promise to pay for future satellite launches. EOSAT was thus given a de facto monopoly on all Landsat images, because there were no direct competitors. The government's policy of providing nondiscriminatory access to remote sensing data on a worldwide basis was interpreted by EOSAT to mean that as long as the company charged the same (high) price to everyone, it was following the government's nondiscriminatory access policy. The price of Landsat images increased from approximately $400 per image to $4,400 per image, a price at which EOSAT was able to attract some commercial and federal government customers, but few academic or independent researchers.

In the early 1990s, the research community became anxious to use Landsat data for global change research, and NASA and NOAA complained to Congress about the high prices. In response to those and other concerns, Congress passed the Land Remote Sensing Policy Act of 1992, which returned the Landsat system to the public sector. NASA and EOSAT negotiated a price reduction to the U.S. Government and Affiliated Users to $425 per image. However, nonfederal and foreign researchers still must pay the high image prices, effectively cutting off a large segment of the research community that would benefit from the use of Landsat images. This situation is expected to be rectified with the launch of Landsat 7, the next satellite in the series, at which time NASA is supposed to make the data available again to all users at the marginal cost of reproduction.

BOX 4.4
The Impact of Landsat Privatization on Research

Forty-five leading scientists were asked to describe the effects that their limited access to Landsat data has had on their work over the past decade.[1] The list below summarizes their most important points:

- Projects in the initial concept often are not proposed, or are drastically scaled down, because the cost of the scenes is prohibitive.
- The development of state land cover maps for many different applied purposes has not been possible.
- Important agricultural inventories were inhibited to the point where efforts were abandoned.
- The high costs of the data significantly hindered studies involving multiple dates and scenes. Developing appropriate and automated methods of change detection has been especially hindered. The conduct of seasonal or phenotypically related research was not possible. Many remote sensing studies have involved analysis of a single scene, and the conclusions derived relate to that (and only that) scene. The repeatability of such conclusions is thus suspect. Would the same observations have been derived using a data set from a different time period? If a given technique does not work with a July data set, should one conclude that the technique does not work at all? Only tests with a multitude of data sets will provide these answers, but high prices limit these rather basic tests.
- One impact of the continuing dilemma with Landsat and privatization has been that the technology is not the state of the art. For more than a decade the United States has had no long-term commitment to land remote sensing at the national level. Landsat 5, the highest-resolution operational U.S. satellite and the only one available for civilian use, is more than 12 years old. The United States has been struggling to develop and launch Landsat 7 (with Landsat 6 having failed on launch). There is now increasing competition from other nations that have committed to long-term operational continuity for their land remote sensing systems and have stepped in to fill this void in U.S. policy.
- Graduate students frequently are restricted in the Landsat data available to them. Some students spend time writing mini-proposals for funds just for acquiring data and these proposals are often rejected because of unavailability of funds.
- The ability to conduct low-cost basic research has been hampered. Many technological innovations have come out of such research, but because of the cost of new imagery, these opportunities have been reduced or the research has been completed with older, archived data. While technology development is somewhat independent of the age of the data, applied research and environmental managers become less interested when such efforts use data long out of date. Even now, there are numerous articles in the journals using data from the 1980s. Techniques such as subpixel classification, research into periodicity patterns as they affect image quality, patterns of spectral mixing in patchy landscapes, and climate/microclimate/image calibration are all significantly under-researched. Thus, the high cost of imagery has resulted in few, if any, innovations in the applications of this technology.
- The poor availability of Landsat Thematic Mapper (TM) imagery is not only due to cost, but also to the practice of operating the satellite selectively for

certain land-surface areas. The cost of imagery reduced the user base, and EOSAT had to determine which images would be most marketable prior to acquiring them. That left many scientists with a very limited, high-cost archive of TM data. Thus, for certain areas of the globe there is extensive coverage, but for others it is very poor.

In addition, the more detailed narrative of a senior soil scientist, a member of the interdisciplinary Laboratory for Applications of Remote Sensing at Purdue University, is revealing:

> During the mid- to late 1970s and into the early 1980s, our research group was heavily involved in interdisciplinary research (involving electrical engineers, civil engineers, computer scientists, statisticians, meteorologists, crop physiologists, soil scientists, foresters, environmental specialists), collaborating with several federal agencies and other universities. Our research focused on the study of the relationships between the data derived from the Landsat multispectral scanner (MSS) and thematic mapper, the characteristics of agricultural land surfaces, and the changes of these surfaces with time. Specific objectives were to determine the feasibility of using Landsat MSS and TM data for crop inventory and monitoring. Some questions addressed were: What quantitative changes occur in the spectral characteristics of crops (corn, soybeans, winter wheat, spring wheat, rice, many other crops) throughout phenological development? What spectral changes are a result of stress from drought, nutrient deficiency, disease, insect infestation, salinity of the soil, storm damage, wetness, and other causes? Can these changes be identified and delineated by classical pattern recognition analysis applied to multispectral data obtained by aerospace sensors? Many of these questions were addressed by our research during that period and at least partially answered.
>
> One of the many areas of research that came to a complete halt when the price of TM data was increased manyfold was in the application of time-sequential remote sensing data (e.g., MSS, TM, advanced very-high-resolution radiometer (AVHRR), and others) to mapping and monitoring terrestrial ecosystems and to developing models to assess land quality, soil productivity and degradation, and erosion hazards. The anticipation that had begun to build for use of earth observation satellite data for integration with other data sets to provide national, continental, and global resource databases was suddenly dashed. It became impossible to develop the procedures, approaches, and models for doing any credible global monitoring and modeling, especially for terrestrial ecosystems, without such data. Remote sensing research with agricultural crops slowed considerably, and much of it stopped completely as a result of the diminished support for civilian space research in the early 1980s and the subsequent commercialization of Landsat, which resulted in exorbitant data costs. Researchers in remote sensing laboratories and centers around the world, especially in universities, almost overnight went from a "data rich" situation to a condition of "data poverty." Many of the basic questions that were being addressed in the early 1980s are still being asked because the data became unaffordable to the research community addressing these questions. The resulting nonavailability of data probably played a significant role in the decline of and, in many cases, the closing of remote sensing programs at numerous universities.
>
> It is a pity that the commercialization occurred when it did. The scientific "homework" had not been completed or carried out to the point at which marketable products had been sufficiently demonstrated. Another few years of affordable data and public research support in this area might have made the commercialization process more feasible and ultimately less painful.

---

[1]Compiled by Michael D. Jennings, National Biological Service, Gap Analysis Program, University of Idaho, Moscow.

However, if not done in accord with the principles above, it can be disastrous for the government's basic scientific mission. Such a situation occurred with the privatization of the Landsat system in 1985. The history of this case and its impact on science are described in Boxes 4.3 and 4.4.

It also should be noted in this context that the government producers of scientific data are themselves in a potential position to act as exploitive monopolies, absent some formal restraints. In the United States, the prices that the federal government can charge are heavily regulated by law (see the following section). This is not the case in all other countries, however; some government agency monopolies are allowed to sell their data commercially without the restraints of a competitive market.

In summary, when privatization or commercialization leads to unregulated monopoly supply, it is not good public policy, and especially not good for science. When privatization leads to competitive supply to multiple user communities, it could well be good public policy, especially if scientific users are assured access at reasonable prices and there is a net benefit to the public from such a transfer.

PRICING PUBLICLY FUNDED SCIENTIFIC DATA

Ramsey pricing is a mechanism developed for regulated monopolies, which in this context mean either government monopolies or monopolies acting as agents of the government. It was long ago proved by the British mathematician, Frank Ramsey,[11] that where the optimal price for a good is zero or is insufficiently high to pay for the total cost of the product, economic efficiency requires the shortfall to be covered by differential prices, with the highest prices charged to users with low elasticity of demand, that is, users whose usage will be reduced relatively little by a given charge for the item. The reason is clear. A high price to a user whose elasticity of demand is great will cause a large cut in that person's use of the item—a major deviation from that individual's optimal use of the item. Thus, any price that is likely to prevent scientists from using data because of budgetary problems means that the elasticity of demand of the scientist users is high. Ramsey analysis then confirms that these scientists should not be charged a substantial price for data. Commercial users, in contrast, if they stand to make a considerable profit from the data, will acquire them even at a substantial price, meaning that their demand is relatively inelastic. Welfare analysis confirms that these users should bear the bulk of the cost.

There are several approaches to achieving so-called "Ramsey" prices. The most straightforward approach is pure price discrimination. The supplier can establish different prices for different customer groups, using some means of identifying which customers are in which groups. For example, professional societies and scientific journals often charge different dues or subscription prices to libraries, to professors of various ranks, and to graduate students. Since student discounts are generally quite deep, some form of identification is required for such discounts. Usually, societies depend on self-reporting to discriminate among full, associate, and assistant professors, or by income. Naturally, there is some "leakage" due to false reporting, sharing of materials, and the like. However, these schemes generally are successful in reaching multiple user groups while maintaining revenues.

In some cases, price discrimination based on observable customer characteristics is not feasible. A useful approach in such cases is for the supplier to develop data product lines, consisting of somewhat different packages of data and value-added services at different prices, directed at different user segments. The price differentials cannot be so great as to encourage excessive leakage of users from one package to another because of price. In order to prevent this, price differentials can be no greater than the value differentials among user segments. There are several methods of product line differentiation:

- *Time.* Users who demand up-to-the-minute (i.e., real-time or near-real-time) data pay more than those willing to wait and use archived or retrospective data.
- *Customer support.* Users who demand full access to telephone support and other types of support services pay more.
- *Sample size.* Users who want a higher sampling rate pay more.
- *Documentation.* Users who require full documentation of the data pay more.
- *Scope of coverage.* Are the data useful for a narrow user group or for a broader, more comprehensive audience?

However, these methods do not address *how much* should be charged to various users of the scientific data. The appropriate guidelines are as follows:

- *For the contributor and active researcher community*, data acquisition should most likely be free, or, following much of current practice, available at the marginal cost of distribution. This guideline is based on the presumption that the individual contributor is actually creating the value. In this case, the data distributor is acting as a repository of the data for the active research community, which is both contributing and using the data.[12] However, it is possible that free access may generate such a great demand for service that congestion occurs, in which case some form of congestion pricing would be appropriate.
- *For others, including commercial users*, data acquisition should be priced to cover the costs of serving those users.

The two appropriate pricing methods, incremental and marginal cost pricing, differ as follows:

- *Incremental cost pricing.* The price to secondary users is set so that revenues cover the cost to provide this incremental use, including recompiling the data, perhaps maintaining a computer site for downloading, purchasing CD-ROM blanks, recording the data, shipping to the user, and customer support, but not including costs for the core service.

- *Marginal cost pricing.* The price to secondary users is set at the marginal cost of the specific unit sent to the user, including the cost of the CD-ROM blanks and postage and shipping. This price is lower than the incremental cost price, as long as the cost of output per unit declines when volume increases. It is easiest to express the difference between these two ideas mathematically.

Assume that the total cost to supply the quantities $q_1\ q_2 \geq 0$ of two goods is $C(q_1, q_2) \geq 0$, with $\frac{\partial C}{\partial q_i} > 0$; then

$$\text{marginal cost of good } 1 = \frac{\partial C}{\partial q_i};$$

$$\text{incremental cost of good } 1 = \frac{C(q_1, q_2) - C(0, q_2)}{q_1}.$$

The pricing policy specified in Office of Management and Budget (OMB) Circular A-130,[13] which applies to all federal government agencies, corresponds to incremental cost pricing. The tradition in the research community, and the pricing level indicated by the "full and open access" policy, corresponds to marginal cost pricing.

One can argue on the basis of public good benefits that the price floor should be zero or the marginal cost. If avoiding undue subsidization were to become an overriding concern, then incremental cost would be the appropriate price floor.[14] Under the OMB pricing policy for federal government data, however, the full incremental cost is also the ceiling.

ELECTRONIC ACCESS AND INTERNET CONGESTION

Perhaps the most profound change associated with the digitization of science is the ability to access scientific data worldwide from the desktop, via the Internet. But new capabilities give rise to new problems. In this case, the recent congestion on the Internet, particularly across the Atlantic and the Pacific, has reduced the ability of scientists to access data around the world, and particularly to monitor experiments overseas in real time. Once the exclusive preserve of scientists, the Internet has attracted so much interest since the advent of the World Wide Web that nonscience traffic now dwarfs scientific traffic. This phenomenon is a classic example of a "tragedy of the commons," in which use of a common property becomes so intense that its value and benefit to its users diminish, possibly even to the extent that the good becomes useless. Many scientists see the congestion resulting from this popularization of the Web as interfering with their ability to do science across the Internet, as discussed in Chapter 2.

The Internet congestion issue is a difficult one and is likely to be with us for a long time. Prior to the popularity of the World Wide Web, the fairly low level of usage imposed on the Internet by scientists was well within the modest capacities of the network to function without discernible delay. Scientific users perceived the network as having no capacity constraints, because they never encountered any. Today, the situation is different; the use of the Internet has increased to the point that those modest capacities, even though they are expanding, are being reached (or even exceeded in peak periods) in many areas. Delays are the norm in some situations, such as the link to Central Europe during daytime hours on either side.

What are those capacities? Generally, capacity constraints exist within computers (servers) and the transmission pipes that connect them. These transmission pipes are leased telephone company lines and share the same physical facilities as all other telecommunications services. For example, there are several physical transmission facilities that span the North Atlantic, the most important of which are the undersea fiber-optic cables. All telephone traffic and leased lines use these facilities, with the split being determined by how many lines have been leased for Internet use.[15] Additionally, local "hot spots" can occur, in which a server that hosts a particularly popular Web site becomes congested because of increased traffic. This in turn causes nearby servers to become congested with traffic attempting to reach the busy server, so network congestion can spread. The solution is straightforward: the popular server must add capacity.

Note that the greatest strength of the Internet—its decentralized character—can become its greatest weakness. In a fully decentralized network, the solution to congestion relies on the implementation of decisions to expand capacity. Unfortunately, congestion affects not only the party that causes the congestion but others as well, so it becomes a classic "externality" problem. The owner of a site may be perfectly happy to live with congestion on his or her popular site (even though the site's visitors are not happy), but the owner's actions will cause congestion at nearby sites and perhaps even throughout the network, thereby imposing costs on others. Furthermore, congestion can also be imposed when two parties cannot come to an agreement about expanding capacity. If the United States and European countries cannot agree on how much capacity is needed to connect their respective networks, or how to share the costs of that capacity, then needed expansion may not occur, to the detriment of users.

How best to deal with this "tragedy of the commons"? The usual solution to congestion[16] is to ensure that those who cause it bear its full cost, in the form of congestion prices. This is a mature theory in economics. The Varian and MacKie-Mason proposals for "smart pricing" are the most well-developed in the context of the Internet and are specifically designed to cope with congestion problems.[17] It is sufficient for the committee's purpose here to note that such pricing schemes will generally involve some form of usage-based prices for real-time traffic (i.e., traffic that cannot be delayed until congestion has eased). This could be a per-packet charge, for example, or a per-minute charge differentiated by type of traffic (e.g., telnet, video, Internet, phone).[18]

The effectiveness of congestion prices depends on two obvious, but critical, functions served by prices: (1) users will have an incentive to postpone traffic to less congested, lower-cost periods if they are required to pay a high price during peak periods, and (2) suppliers will have an incentive to increase the capacity of a server or a transmission route if traffic during the peak periods is highly profitable. The demand-shift effect tends to reduce traffic, while the supply effect tends to increase capacity in the long run. Additionally, users who require peak-load use in the short run will, upon payment of the peak-load price, be more likely to obtain service under this regime than with underpriced peak service.

One other possible means for dealing with congestion would involve investment in hardware and especially software and would follow a growing trend in business. This would be the creation of one or more dedicated networks for scientific research, such as the Internet II now being developed with support from the National Science Foundation. Such a network would function much like the "intranets" being established by private firms. The committee urges that funding agencies and professional societies begin to examine and evaluate this option in greater detail.

There is no question that congestion is also its own punishment. Servers or network administrators who generate congestion also suffer its consequences. However, those consequences extend to others, and the individual disincentives will not reflect the "external" costs imposed. Current administrative mechanisms may help alleviate congestion, as well. Sufficient peer pressure from network "partners" may induce managers of congestion-causing servers to increase their capacity to keep their standing with their colleagues. In the long run, however, the "partnership" model of the past is unlikely to provide sufficient incentives to alleviate congestion, and the current situation can be interpreted as a transition to a new regime in which more formal mechanisms, such as congestion pricing, will be required.

RECOMMENDATIONS REGARDING ECONOMIC ASPECTS OF SCIENTIFIC DATA

The committee recommends that the economic aspects of facilities for storage and distribution of scientific data generated by publicly funded research be evaluated according to the following criteria:

- *Does the scientific research depend on a substantial public investment in one or more facilities that generate the data of interest?* If so, the data distribution facilities are most likely to benefit by being vertically integrated with the observational or experimental facilities themselves.

• *Does the (non-facilities-based) distributed scientific research involve coordination among researchers, possibly in different countries?* If so, then data distribution becomes a means of communication among contributing scientists, and for this community, the price of the data alone should be zero. If the distributor subsequently adds value to the data, then the price should be no higher than the marginal cost of adding value.[19]

• *Is the community of users roughly the same as the community of contributors?* If so, then data distribution should be priced at zero (or at marginal cost, if value is added). If there are many users who are not contributors, such as commercial customers, then some form of price discrimination to ensure zero or low prices to contributing scientific users, with possibly higher prices to others, may be appropriate.

• *Is the user community large enough to support more than one data distributor?* If so, then privatization of data distribution may be a viable policy option. If not, then privatization should occur only if the contractual arrangements are adequately protective of the needs of the scientific community. Necessary—but not necessarily sufficient—conditions for privatization to be desirable are as follows:

—The distribution of data can be separated easily from their generation.

—The scientific data set is used by others beyond the research community.

—It is easy to price discriminate/product differentiate between scientific users and other users, and it is easy for the government to contractually mandate low prices to scientific users for government-funded data.

—Privatization will not result in the unrestricted monopoly provision of the data.

The appropriate price ceiling for nonscientific users of scientific data generated through government research is incremental cost, as defined in the section above titled "Pricing Publicly Funded Scientific Data." The price of scientific data to the contributing scientific community should be zero, or at most marginal cost.

NOTES

1. Not all scientific data are maintained and distributed by public agencies or via public funding, although this is the norm. Various not-for-profit institutions and private firms are also providers of basic scientific data.

2. Throughout this section, the terms "scientists" and "scientific researchers" are intended to include explicitly both U.S. and foreign scientists working in the natural sciences. Scientific research is by its nature a global enterprise; actions by any single government are felt by the entire scientific community. However, the barriers to scientific information flow are exacerbated by the problem of unequal technological capabilities, important in virtually all the disci-

pline-specific contexts. This has at least two dimensions: individuals/institutions have different levels of computing and networking skills, and individuals/institutions have different levels of accessibility to hardware and software needed to exercise those skills. Generally, scientists are relatively sophisticated in their skill sets; the problem in science is the availability of tools, both computing and networking. The digitization of science suggests that if individuals, institutions, or countries lack the tools to process, analyze, and share scientific data at the level typically enjoyed in the developed countries, they will be unable to participate fully in the scientific endeavor, to the detriment of science as a whole. The problem, of course, is not confined to science, and involves the unequal distribution of wealth and income across the globe. At this level, it is highly unlikely that the expressed needs of scientists will have much effect in changing this unequal distribution of income. However, specific science-oriented activities can make a difference; recycling hardware, for example, to scientists in developing countries may greatly improve their ability to process data at almost no cost. While U.S. scientists may view two-year-old 486 PCs as hopelessly underpowered, these cast-off computers may be a godsend to scientists in Africa who currently have nothing. Perhaps the best way to facilitate such transfers is via professional society programs and institutions, which are likely to be able to identify both needs and donors efficiently, with relatively low levels of public financial support.
3. The external benefits of fundamental or basic research are to be contrasted with the benefits of development, generally of products and services for sale, in which the full benefits are captured by the buyer and seller. Generally, only basic research is acknowledged to be a public good, while development is seen as a private good, at least where there is effective patent protection. Of course, the distinction between research and development is not as clean as this suggests, although it is often a useful distinction.
4. Of course, the committee recognizes that the acquisition of data by the government through its research activities is neither a costless activity nor an activity requiring unfettered spending; it is, however, a part of the process of doing research that falls outside the charge of this study, which focuses on data distribution and access.
5. An instructive lesson can be drawn from the shift in public policy regarding mental health in the 1960s. The strategy was to shift from institutional care to community-based care, with substantial deinstitutionalization of patients and a funding shift to community facilities. The unfortunate result was that funding for institutions dropped, but no funding was provided for community care. It is claimed that this failure to provide for local funding has added substantially to the homeless population.
6. A more extensive introduction to the complex topic of price discrimination/product differentiation is contained in the section "Pricing Publicly Funded Scientific Data," below.
7. The committee's use of the term "private" in this context includes only for-profit firms that are not subsidized by the government; other private institutions such as universities, educational consortia, foundations, and other NGOs are not included in this context.
8. As a practical matter, one would expect scale economies of data distribution to be exhausted at rather low levels of demand. However, many scientific data sets may have demand lower than this threshold and thus be subject to a "natural monopoly."
9. In this case, of course, meteorologists still obtain their scientific data from NOAA, not commercial firms, and so there are actually separate distribution channels for scientists and the public. The point here is that there are activities in which the private sector can outperform the public/NGO sector.
10. Implicit is the assumption that if the government and private suppliers (who may get their raw data "wholesale" from the government) are competing against one another in the marketplace, the government is constrained to set prices to recover its cost. See the section titled "Pricing Publicly Funded Scientific Data."
11. F. Ramsey (1927), "A Contribution to the Theory of Taxation," *Econ. J.*, 35 (March):47-61.

See also M. Boiteaux (1956), "Sur la gestion des Monopoles Publics astreints a l'equilibre budgetaire," *Econometrica* (Jan. 24): 22-40, and W.J. Baumol and D.F. Bradford (1970), "Optimal Departures from Marginal Cost Pricing," *Am. Econ. Rev.*, 60 (June):265-283.

12. A zero price might seem too low to some; however, it is the contributions of this community that actually create the value in the first place. An instructive analogy is consumer banking: depositors with sufficiently large balances receive "free" checking from banks; the *quid pro quo* is that the bank has the use of their money. Similarly, the database provider could give free access to contributors, with the *quid pro quo* being the contributions themselves. It also should be noted that if the data were not supplied free, it is almost certain that some public-spirited scientist with spare server capacity would have a graduate student maintain an FTP or WWW site with the data on it for free downloading for interested colleagues.

13. The OMB Circular A-130 policy regarding federal government information dissemination practices was codified in the Paperwork Reduction Act of 1995, P.L. 104-13, which amended 44 U.S.C. Chapter 35, effective October 1, 1995.

14. The original reference for this finding is Gerald Faulhaber (1975), "Cross Subsidization: Pricing in Public Enterprises," *Am. Econ. Rev.*, 65:966-977. Later references, among many others, are J.C. Panzar and R.D. Willig (1977), "Free Entry and the Sustainability of Natural Monopoly," *Bell J. Econ.*, 8 (Spring):1-22, and W.J. Baumol (1977), "On the Proper Cost Tests for Natural Monopoly in a Multiproduct Industry," *Am. Econ. Rev.*, 67 (December):809-822.

15. The current situation on the North Atlantic route is that there is virtually no congestion for placing telephone calls from Europe to the United States, but serious congestion for Internet traffic. The conclusion from this evidence is clear: there is plenty of capacity in the physical transmission facilities, but too little of that capacity is devoted to the Internet. The simple (but expensive) solution for the researcher in Europe is to place a modem telephone call to the U.S. computer, or vice versa, and conduct the research via direct connection.

16. There are, of course, a wide variety of engineering and queuing-theory solutions, priority schemes, compression methods, and so forth. All such methods either reduce the capacity required or seek to allocate scarce capacity during congestion to more valued uses. Ultimately, however, capacity is finite, and congestion may still occur, but at higher load levels.

17. For a discussion of the economics of network pricing, see Hal Varian and Jeff MackieMason (1995), "Pricing the Internet," in B. Kahin and J. Keller, eds., *Public Access to the Internet*, MIT Press, Cambridge, Mass.; Hal Varian and Jeff MackieMason (1995), "Pricing Congestible Network Resources," Advances in the Fundamentals of Networking, *IEEE Journal on Selected Areas in Communications*, and Gerald R. Faulhaber (1992), "Pricing Internet: The Efficient Subsidy," in B. Kahin, ed., *Building Information Infrastructure*, McGraw-Hill, New York.

18. Another possibility for dealing with congestion is to offer a lower subscription charge to users who are willing to postpone their use to off-peak times. Currently, Lexis/Nexis offers universities a low subscription charge but denies access during peak times.

19. By "adding value" in this case is meant any transformation of the data beyond that necessary for scientific research that increases the value of the information for some or all potential users of the data.

5

The Trend Toward Strengthened Intellectual Property Rights: A Potential Threat to Public-Good Uses of Scientific Data

Laws and regulations, both national and international, affect the flow of scientific information through electronic networks. Among them are rules regarding liability for false or misleading information, laws protecting individual privacy rights, and export controls. This chapter focuses on intellectual property policies and their expression in laws and regulations affecting the contents of databases, because changes are now afoot that may erode the relatively privileged position science has held within the existing legal framework.[1] By restricting scientists' full and open access to the data on which future advances depend (see Box 5.1), these changes could impede the progress of science and thus limit the contributions that science can make to society, notwithstanding the constitutional mandate that intellectual property rights should be limited in time and should advance science and the useful arts.[2]

One such change is that governments, including our own, are finding it increasingly difficult to maintain the rate of growth that publicly funded science has enjoyed over the past half century. When scientific research is supported instead by private funding, the end results and perhaps the research itself are likely to be kept proprietary. Furthermore, there are indications that the scientific data management that governments continue to fund may well be carried out as if it were proprietary, in the sense that fees for use of the data may exceed the costs of dissemination. As their tax bases decline and governments come to regard their data collections as possible sources of revenue, they have, in some instances, adopted the same short-term, profit-maximizing strategies as private firms. Yet today, when commercially valuable data of scientific importance are made available in electronic form, they also become available for rapid, inexpen-

> **BOX 5.1**
> **Effects of Government Support on U.S. Research and Data Activities**
>
> The ability of private-sector technological development in the United States to thrive without the kind of centrally organized institutional framework and industrial policy apparatus typical of the European Community and many other nations[1] has stemmed in good measure from the large public investments in basic research and development that were made after the late 1950s, in response to Cold War pressures and national security interests. In retrospect, the success of the U.S. innovation system, despite its apparent anarchical character, can be seen as linked to public funding of academic institutions and specialized laboratories, whose research product has paved the way for private industrial applications.[2] In this context, the fact that federal funding also largely defrayed the costs of collecting and disseminating raw and elaborated scientific data merits particular attention.
>
> Throughout the Cold War period, and extending into the present, the U.S. government has reinforced its subsidies of fundamental research with a policy of open exchange of scientific data. This policy was promoted internationally through the government's bilateral science and technology cooperative agreements and increasingly in recent years through both bilateral and multilateral agreements concerning various large-scale research programs and projects. None of these agreements, however, has broadly encompassed all scientific research activities. Instead, they typically have been limited to scientific cooperation and related protocols for the exchange of data according to the special interests of a geographic region, scientific discipline or subdiscipline, or specific projects undertaken by the parties to the agreement. In the rise of the United States to become the world's leading producer of technological goods and scientific information, the government's role in ensuring an open supply of data to the scientific community under favorable economic conditions has been a constant stabilizing factor.
>
> ---
>
> [1]See, e.g., Margaret Sharp and Keith Pavitt (1993), "Technology Policy in the 1990s: Old Trends and New Realities," *J. Common Mkt. Stud.*, 31:129, 138-39, table 1.
> [2]See Computer Science and Telecommunications Board, National Research Council (1995), *Evolving the High Performance Computing and Communications Initiative to Support the Nation's Infrastructure*, National Academy Press, Washington, D.C.

sive copying and manipulation. While this facilitates value-adding uses from one perspective, from another, it undermines the data provider's ability to recover costs, much less to generate a profit.

A second change is that, in many areas of research, the separation has diminished between basic research, where intellectual property rules are more concerned with attribution of ideas and findings than with the appropriation of published material, and applied research, where intellectual property and proprietary concerns predominate. This conjunction has been especially evident in computer science and biotechnology, where some basic advances are now virtually inseparable from their industrial applications. The granting of patents or other exclusive property rights in these industrial applications can affect the ability of other researchers to test and extend the theories underlying them.

Third, the revolutionary convergence of digital, computing, and telecommunications technologies has profoundly altered the preexisting status quo.[3] The potentially large gains and losses from the commercial exploitation of data under these changing conditions have led to a concerted drive for new and stronger forms of legal protection for publishers of electronic databases in general, including compilations of scientific data that were heretofore treated as components of the public domain.[4]

The current trend toward stronger and more enduring intellectual property rights, and fewer limitations on the rights of copyright holders vis-à-vis public-good uses of information, could reduce some of the limitations that have benefited scientists, and on which they have relied. Government studies of the challenges that digital technologies pose for intellectual property law at both national and international levels have stimulated calls for strengthening intellectual property rules. In addition to legislation either adopted or still under consideration in the United States and other nations, proposals to strengthen international copyright and related laws were a major focus of multilateral negotiations sponsored by the World Intellectual Property Organization (WIPO) in December 1996, largely at the urging of the United States and the European Union.

One of the draft treaties currently being considered calls for worldwide adoption of a new form of intellectual property protection for the contents of databases.[5] Although this treaty was scheduled for discussion and approval at the WIPO Diplomatic Conference held in Geneva, Switzerland, on December 2-20, 1996, the conference delegates decided that it required further study. Future adoption of a treaty with similar proposals would have such profound consequences for transnational exchanges of scientific data that the committee chose it as a principal focus of this chapter.

This chapter begins by briefly describing the relevant legal infrastructure during the predigital period and by identifying certain destabilizing factors, such as the introduction of electronic photocopying machines. It then outlines digital technologies' role in accelerating the rise of information as a commodity to be bought and sold and in thus upsetting the previous imperfect balance between underprotection and overprotection of the rights of data creators and holders. The discussion that follows examines the emerging legal responses to these phenomena. It describes the current legislative and treaty proposals in detail and explores the implications for science of new proprietary rights in databases. The chapter concludes by proposing actions that groups representing the research and education communities should undertake to stimulate reformulation of the legislative and treaty proposals, with a view to reconciling the need to protect the legitimate interests of database makers with the need to protect the activity of science and to ensure its ongoing contribution to the public interest.

SALIENT FEATURES OF THE PREDIGITAL STATUS QUO

Because the creation, collection, and dissemination of scientific data in the United States have in large part been subsidized by government funding, they have not depended on the balance between incentives to create and efforts to preserve free competition that intellectual property law normally governs. Within this framework, most academic compilers or generators of scientific data were more concerned about obtaining credit or recognition for their contributions than about securing the economic fruits of their efforts.[6] Only in cases where members of the scientific community authored discursive scientific works or otherwise participated in applied technological innovation, or where commercial publishers compiled value-added databases, were they likely to be affected by legal rules governing commercial applications of data. In such cases, existing legal institutions proved relatively stable in the predigital epoch, and the scientific community has taken this stability largely for granted.

In the private sector, by contrast, commercial compilers of data have long suffered from a risk of market failure owing to the intangible, ubiquitous, and, above all, indivisible nature of information goods and to the ease with which free riders may have appropriated the fruits of the compilers' investment, once the information goods were made available to the public in print media. Despite this risk, the domestic and international intellectual property systems responded laconically, if not with indifference, to the compilers' dilemma.[7] This indifference stemmed in part from the inability of the worldwide intellectual property system to match compilations of data to the basic subject matter categories covered, respectively, by the Paris Convention for the Protection of Industrial Property (1883) and the Berne Convention for the Protection of Literary and Artistic Works (1886).[8] It also stemmed from a concomitant reluctance to fetter the basic building blocks of scientific and intellectual discourse with legal impediments.[9]

Notwithstanding these infirmities, the commercial exploitation of nonscientific data and of published compilations of information prospered in some developed countries, notably the United States and the United Kingdom (where copyright protection is sometimes available). Whether there would have been greater commercial exploitation of scientific data in the past if publishers could have invoked stronger proprietary rights is a matter of conjecture. Patents are seldom available for database contents because writings are not patentable subject matter and also because the largely incremental character of database development would typically make it hard to meet eligibility requirements. Even contract law has significant limitations when mass-market information products are sold to persons outside the scope of a contract.[10]

Data as Know-how Applied to Industry

To obtain some measure of protection, firms engaged in industrial applications of scientific discoveries have entrusted their commercially valuable data to trade secret law or to equivalent laws of confidential information. Trade secret law (or equivalent laws of confidential information) provides innovators and investors with no exclusive property rights at all. Rather, it permits third parties to reverse-engineer any unpatented industrial product by proper or honest means in order to reveal the process by which it was obtained, and to use that process to manufacture equivalent goods.[11] To the extent that an innovative product is derived from commercial applications of scientific data kept under actual or legal secrecy, a competitor always remained free to generate the same data and to apply it to similar products or uses.[12]

Trade secret law thus provides qualifying originators with no legal immunity from direct competition. It merely confers a "head start," that is, an uncertain period of natural lead time, during which originators seek to recoup their investment in research and development while establishing their trademarks as symbols of quality that consumers recognize. In this and other respects, trade secret law operates as a liability regime that discourages certain types of conduct rather than as an exclusive property right that may create a legal barrier to entry (see Box 5.2).

When scientific data are disseminated to the public in print media, they normally forfeit the protection of trade secret law, or related laws of confidentiality, except insofar as two-party contracts may otherwise provide. Not surprisingly, commercial compilers in such cases have sometimes found it difficult to appropriate the fruits of their investment unless either copyright laws or unfair competition laws afford them a limited shelter against wholesale duplication by third parties.

Copyright Law as a Cultural Bargain

The advent of the printing press created for published literary and artistic works markets that had previously existed only in a rudimentary form, owing to the need to produce each copy of a work by hand from a single original. Paradoxically, to promote markets for information goods and other literary and artistic works, the state intervened by erecting new monopoly rights—intellectual property rights—even as it removed the royal privileges and guild monopolies pertaining to tangible goods that were handed down from the Middle Ages.

Information goods have the properties of so-called public goods—they are nondepletable and nonexcludable. A second comer's use of a new information good does not diminish or exhaust it; once it is disclosed to the world, anyone can use it without the originator's permission and without reimbursing him or her for the costs of research and production. Unless the state limits the ability of third parties to copy a given literary production, for example, and to sell the copied

> **BOX 5.2
> Definitions**
>
> **Liability Rule and Exclusive Property Rights**
> A property right precludes third parties from appropriating the object of protection, whereas a liability rule regulates the means by which they can engage in certain potentially harmful acts on certain conditions.[1] If one has "rightful possession of some thing—such as an automobile or a home" under an exclusive property right, "another person ordinarily cannot take it without permission"; but a liability rule permits others to engage in acts that "create risks of harm and thus constitute probabilistic invasions of property interests" (such as nuisances), while obligating them to pay damages for harm under specified circumstances.[2]
>
> **Sui generis**
> Sui generis means "of its own kind or class" (*Black's Law Dictionary 1434*, West, 6th ed., 1990). The literature refers to special-purpose intellectual property laws that deviate significantly from the classic patent and copyright paradigms as "*sui generis*" regimes. See, for example, Pamela Samuelson (1985), "Creating a New Kind of Intellectual Property Law: Applying the Lessons of the Chip Law to Computer Programs," *Minn. L. Rev.*, 70:471 (discussing the *sui generis* character of the Semiconductor Chip Protection Act).
>
> **Subpatentable**
> A subpatentable innovation is novel in the sense of being new, but it represents a step in technical progress that an engineer might be expected to make in due course. By definition, a patentable invention must be "nonobvious" in the sense that it represents a breakthrough beyond the capacity of a routine engineer to make in due course.[3]
> In simpler terms, patents are supposed to reward extraordinary achievements, while subpatentable innovations are those that proceed in small, incremental steps.
>
> ---
>
> [1]See, e.g., Guido Calabresi and A. Douglas Melamed (1972), "Property Rules, Liability Rules, and Inalienability: One View of the Cathedral," *Harv. L. Rev.*, 85:1089.
> [2]Louis Kaplow and Steven Shavell (1996), "Property Rules Versus Liability Rules: An Economic Analysis," *Harv. L. Rev.*, 109:713, 713-15. For an analysis of trade secret law as a default liability regime governing relations between originators and borrowers of subpatentable innovations, see Reichman, "Legal Hybrids," note 27.
> [3]See 35 United States Code, section 103 (1996).

good for less than the price charged by the originator, neither the author nor the publisher may have sufficient incentives to create or invest in the dissemination of cultural and information goods.[13]

The historical solution to this problem has been the mature copyright system, which charges both authors and their publishers a price for overcoming market failure. In effect, copyright law has enabled the state to impose "portable fences" that accompany intangible creations and that limit what purchasers can do with them, even though they possess the physical artifacts, such as books or printed tables of numbers, in which these intangible creations are embodied.[14] In so doing, the state also has imposed legal constraints on authors and publishers—a cultural bargain—that has attempted to balance incentives to create against the public interest in both free competition and access to the copyrighted culture.[15]

For example, although copyright law protects an author's personal expression for a relatively long period of time, it attaches only to "original works of authorship." In principle, this requirement excludes functionally dictated collections of data that fail to manifest a creative selection or arrangement.[16] Moreover, copyright law never prevents third parties from independently creating their own versions of another author's unprotectable ideas or of the factual discoveries presented in a given scientific publication. In other words, copyright law protects only a given author's style, not his or her factual or ideological content.

The Concept of Fair Use

The mature copyright paradigm further curbs even this limited monopoly by relaxing the author's control over certain uses of great public interest. Thus, numerous exceptions to and limitations on the copyright owner's bundle of exclusive rights favor face-to-face teaching (e.g., by allowing limited duplication of materials for classroom use), library and archival uses, and selected public interest pursuits,[17] in addition to a general "fair use" exception "for purposes such as criticism, comment, news reporting, teaching . . . scholarship, or research."[18]

While the availability and scope of statutory exceptions usually vary with the nature of the subject matter at issue, the fair use exception applies to all subject matter categories across the board. Even so, overriding the copyright owner's exclusive rights in the name of fair use remains an atypical result contingent on a judicial evaluation of the special "purpose and character" of the use, the "nature of the copyrighted work," the "amount and substantiality of the portion used," and the "effect of the use upon the potential market for or value of the copyrighted work."[19]

In recent years, the advent of new technologies—from photocopying machines to computer programs and optical scanners—has unsettled the doctrine of fair use[20] by enabling even copies for private research uses to displace commercial markets,[21] and also by making it possible to overcome most of the transaction cost problems that increasingly had been used to justify application of the fair use exception in practice.[22]

Protection Afforded

Copyright law will not protect the product of a compiler's industrious efforts—i.e., of labor, skill, or investment—if the selection or arrangement it embodies does not rise to the level of an original work of authorship. Moreover, a mature copyright system usually affords protection only against wholesale copying of the original selection and arrangement underlying any eligible compilation of data. In the United States, this doctrine of weak or "thin" protection for factual works has been reinforced by First Amendment concerns, which some courts and commentators viewed as mandating broad access to the disparate facts that result from a compiler's efforts.[23] When these doctrines apply, they greatly diminish the value of copyright protection even to database publishers who satisfy the eligibility criteria, because their exclusive reproduction and derivative work rights—as construed by the Supreme Court in *Feist Publications, Inc. v. Rural Telephone Service Co.*—will not normally prevent unauthorized extractions of disparate data for either competing or value-adding uses.[24]

Some federal appellate courts, however, have begun to rebel against the *Feist* decision and to reinstate stronger copyright protection for factual compilations and databases by subtle doctrinal manipulation.[25] Whether state or federal unfair competition laws could also provide some supplementary relief against the unauthorized copying of commercially valuable data that are not protected by trade secret or copyright laws remains an unsettled question, although such laws are sometimes invoked both here and abroad.[26] In any event, this cyclical fluctuation between states of underprotection and overprotection is a characteristic trait of borderline subject matter that fits imperfectly within the classical patent and copyright paradigms, such as the contents of databases.[27]

DIGITAL TECHNOLOGY—DISRUPTING THE BALANCE OF PUBLIC AND PRIVATE INTERESTS

Despite (or perhaps because of) the relatively weak legal infrastructure governing use of data, a thriving market for compiled information has grown up, and U.S. publishers appear to play a dominant role in it,[28] although it is important to emphasize that this market has been largely concerned with nonscientific data and information. This industry seems largely characterized by niche marketers who supply and dominate specific market segments. The limited size of these segments and the relatively high startup and servicing costs seem to deter second comers from readily entering such markets.[29] In other words, once the threshold level of investment has been crossed, the first comer tends to take the relevant market segment as a whole.

The public sector nonetheless has remained largely immunized from the potential abuses of market power inherent in this situation, owing both to its subsidized status and to the long-standing legal tradition that denied copyright protection to works produced by U.S. government agencies.[30] As a result, data provided by federally funded projects have flowed through the domestic innovation system with few legal impediments (see Box 5.1), and legal disputes about ownership or the exercise of proprietary rights in scientific data as such rarely have been ventilated before intellectual property tribunals.[31]

By the late 1980s, however, digital technologies and new telecommunications networks had combined to produce "the greatest changes in the way information is distributed since the invention of printing by movable type in the 15th century."[32] The use of computers made it economically feasible to collect, store, manage, and deliver huge amounts of data at a time when continuously expanding databases have become ever more prominent building blocks of knowledge, especially in the observational sciences, as discussed in Chapter 3. Electronic databases further blur the line between these collection and application functions by allowing users to make their own tailor-made extractions from the mass of data available in the collection as a whole.[33] These tools allow users to "add ... immense value to what would otherwise be masses of incoherent, disparate data."[34]

Moreover, the latest value-added data products, once disseminated worldwide via the Internet and other media, frequently lead to the rapid production of new technical innovations, which result in the generation of more data.[35] Electronic publishing thus broadly advances the revolutionary process that computerization began, and it makes both data and research results potentially available at very low cost all over the world.[36]

As this digital and telecommunications revolution has created vast new markets for electronic information goods and tools,[37] it has outpaced the legal infrastructure, which remains geared to the slower-moving print media.[38] This strain manifests itself in two contradictory ways. Sometimes digital technology aggravates the basic market-failure characteristic of information goods and thus deepens a chronic state of underprotection. This can occur, for example, when second comers download the originator's data and enter the market with a competing product that free-rides on the originator's investment.[39] At other times, however, digital technology so thoroughly overcomes the threat of market failure that it endows the first to invest with abnormal market power that can result in a chronic state of overprotection. This can occur, for example, when sole-source data providers charge exorbitant prices or oblige libraries and research institutions to accept terms and conditions that effectively waive both the special privileges and the fair use exceptions set out in the Copyright Act of 1976.[40]

The Vulnerability of Publicly Distributed Electronic Databases

To the extent that government- or university-generated data remain uncommercialized, their vulnerability to technically refined means of accessing, downloading, or duplication is only of relative importance. Presumably, the originators want the broadest possible distribution of their data sets.[41] Even in this situation, however, there are some concerns that are likely to grow over time. For example, government may impose cost-recovery conditions on the use of data that third parties who obtain unauthorized access could avoid. Users also might inadvertently corrupt the original database and cause potential harm.

Moreover, over time, the distinction between basic, noncommercialized data and data applied to industrial pursuits or other downstream uses seems likely to break down, as has already occurred in other disciplines, such as the Earth sciences and biotechnology.[42] Universities and other research institutions may view data compilations generated in the course of their research as potential revenue sources, especially in an era of declining government support, just as they have done with patentable inventions. As more scientific data are applied to commercial purposes for one reason or another, the data collectors must necessarily distinguish between sources that are made publicly available without charge and those that are not.[43] Otherwise, even the providers that do not charge for data could disrupt contractually controlled applications of their own data downstream, not to mention the risk that the noncharging government or academic generator might inadvertently infringe on third parties' proprietary domains.

A related trend is for some governments to commercialize their data, regardless of whether other governments follow suit. The former will become concerned about the vulnerability of their data even if the latter are not. By the same token, those providers that still choose not to charge for their services will increasingly come into contact with (and, perhaps, conflict with) the legal and technical fences that states bent on commercializing data may erect. As one observer put it, "The division between the two regimes" could become "a dam over which information will not easily flow," to the possible detriment of scientific progress and global economic growth, which seems to require that "[m]ore than perhaps any other commodity, data must be allowed to move without barriers."[44]

To the extent that databases are commercialized, whatever their origin, the refined digital technologies that enhance the compiler's power to collect and disseminate data will enhance as well the free-riding competitor's power to appropriate the fruits of the first comer's investment.[45] The second comer who purchases the originator's product, say, in the form of a CD-ROM, may electronically extract and recompile the data in question at a fraction of their collection and distribution costs. The second product may then be sold for less than the first, because its publisher has contributed nothing directly or indirectly to the research and production costs. Digital technology also enables second comers to extract and recombine the originator's data into value-added products that improve on the original, or that compete in different and sometimes distant market segments.[46] In some cases, third parties may even extract the compiler's data in order to make them available over telecommunications networks, an act that can destroy any residual incentives to invest.[47] In such cases, existing copyright laws generally afford little or no relief, as explained above.

Relative Invulnerability of Many Privately Controlled Databases

When the database maker is the sole source of the data in question, and substitute databases cannot readily be compiled from public domain sources, digital technology greatly strengthens a supplier's market power. By restricting access to identifiable, on-line subscribers, for example, and by "placing conditions on access and [using technology] to monitor . . . customer usage," the publisher can largely restore the power of the two-party contractual deal that the advent of the printing press had appeared to destroy.[48] In effect, publishers in this position may not need copyright law at all, even if they qualified for protection. They may prefer to reject the state-imposed cultural bargain in order to override both its fair-use provisions and its specific exemptions favoring the public interest in teaching and research.[49]

Moreover, electronic publishers may have virtually no transaction cost problems to overcome because digital technology now enables them to track and charge for every instance of electronic access, even for browsing and scientific uses that were previously exempt.[50] The resulting market power then enables the publisher to impose monopoly prices and arbitrary terms on users—including libraries, educational institutions, and research centers—and to disregard the social consequences that ensue from the inability of such public organizations to foot the bills.[51]

How Will the Public Interest Be Served in the Information Age?

While many types of scientific data, like other forms of information, possess economic value under the appropriate circumstances, the sponsors of new proprietary rights explicitly contemplate a level of systematic commercialization of both large and small units of data that is unprecedented. How these impending changes in the legal infrastructure will impinge on the research and educational communities has not been clearly worked out even by the European authorities responsible for the European Union's recently adopted Directive on the Legal Protection of Databases.[52] A bill to enact a U.S. model of the European law, which was recently introduced, is even more cryptic in this regard,[53] while the WIPO Draft Database Treaty tried to finesse the issue.[54] One can predict, nevertheless, that these legislative initiatives will greatly affect the scientific and educational communities if, as Chapters 3 and 4 of this report have emphasized, they lead to a more market-driven environment with fewer government subsidies than before.

Whether contractual attempts to reduce users' access to scientific and cultural products that was promoted by copyright laws in the past will survive legal challenges on such grounds as federal preemption of state law, or doctrines of misuse of copyrights (allied to antitrust law), remains controversial.[55] Another question is whether the economic and cultural bargain embodied in copyright law remains appropriate for the digital environment (see Box 5.3), given that trade-driven economic policies have otherwise weakened the consensus on which that

BOX 5.3
Copyright Law in the Information Age—Two Perspectives

The application of domestic copyright laws to digital technologies that did not exist at the time of drafting undoubtedly means that the statutory language will not always fit the cyberspace dimension. The resulting ambiguities could give rise to legal uncertainty in a number of real or hypothetical situations. Yet, the Copyright Act of 1976 was deliberately drafted with a view to accommodating future technologies, while the manner in which one seeks to resolve ambiguities that impartial observers find genuinely troublesome, rather than merely pretextual, depends on one's allegiance to the "economic and cultural bargain" thought to lie at the core of prior law.[1]

For example, bookstores do not charge customers for the privilege of browsing through the stacks prior to purchasing specific books, and this use would not violate the copyright owner's exclusive reproduction rights, irrespective of any contractual relations between booksellers and customers. Once the work in question was converted to digital form and transmitted over telecommunications networks, however, publishers could monitor and charge for analogous uses if they fell within their exclusive rights to reproduce, adapt, publicly perform, distribute, or display copyrighted works, or if a new "exclusive right of transmission" were enacted, as the Information Infrastructure Task Force's White Paper proposes. Enactment of the White Paper's proposed "transmission" right could then explicitly or implicitly remove uses analogous to "browsing" from the prior "fair use" tradition, whereas reliance on existing law would leave the issue to case-by-case judicial determination.

However, the WIPO Copyright Treaty that was opened for signature on December 20, 1996, created in Article 8 a new right of communication to the public that gives authors the exclusive right to make available "to the public . . . [these] works in such a way that members of the public may access these works from a place and at a time individually chosen by them." Moreover, the agreed statements concerning this treaty explicitly carry over the preexisting fair uses recognized in state practice and allow for new instances of fair use for digital transmissions of copyrighted works over telecommunication networks.

The Copyright Act of 1976 also allows one who purchases copies of most copyrighted literary works to give or lend them to others or even to resell them secondhand for profit without owing additional royalties to the owners of the copyrights in question.[2] Whether analogous acts are, or should be, permitted with regard to digitally transmitted works is open to question,[3] as is the right of certain users to

---

[1] See, e.g., 17 U.S.C. §§101 (definition of "literary works" and "copies"), 102(a), (b), 107 (codification of fair use in terms of market interest); Jaszi, note 15 in text (stressing "cultural bargain" underlying copyrighted tradition).

[2] See 17 U.S.C. §§106 (3) (exclusive distribution right), 109 (a) (first sale doctrine). However, unauthorized commercial lending of sound recordings and computer programs is prohibited in the United States, see 17 U.S.C. §109 (b) (1994), and the commercial lending of books or films is subject to royalties under some foreign laws.

[3] See, e.g., White Paper, note 38 in text, at p. 92 (claiming that unauthorized distribution by transmission necessarily entails electronic "copying" (not true of the print media) and therefore violates the reproduction right, which is not limited by first sale doctrine's toleration of lending and resale); McManis, note 64 in text, at pp. 269-72 (criticizing and rejecting the White Paper's interpretation of current law).

(continues)

BOX 5.3 Continued

study or download digitally transmitted works for private research and other noncommercial purposes without payment of royalties to the copyright owners. While publishers of print media had more or less grudgingly come to terms with such private or fair uses, they do not believe the same exceptions apply (or ought to apply) to on-line or other forms of digital transmission.[4]

While the advent of new technologies has always created a degree of legal uncertainty in intellectual property law, the tendency in the past was to allow the law slowly to catch up, despite the risk of some short-term obsolescence. In contrast, the supporters of the currently proposed changes to the law contend that the opposite course of action is needed with respect to digitally conveyed knowledge and information goods, and the major reforms set out in the White Paper adhere to this view.

If one believes that the federal courts can apply existing copyright and unfair competition laws to the new technologies with relatively little friction, then one has implicitly opted for a wait-and-see approach or at least for a minimalist approach, based on case-by-case judicial decisions and a minimum amount of tinkering with the statute as it stands.[5] This approach leaves the traditional exemptions for scientific and educational users intact, but subject to case-by-case evaluation.[6] If, in contrast, one believes that gaps in the law leave on-line publishers too much at risk, then proposals for statutory reform easily escalate into a campaign to rid the emerging information infrastructure of allegedly anachronistic vestiges of the cultural bargain that had heretofore protected users and second comers of works in print and other media.[7]

---

[4]See, e.g., White Paper, note 38 in text, at pp. 65-68, 216; McManis, note 64, at pp. 263-73 and note 258 (criticizing this position and identifying White Paper with the view that "given the availability of metering [by commercial digital content providers], systematic electronic browsing by academics should not be considered fair use"); Pamela Samuelson (1994), "Legally Speaking: The NII Intellectual Property Report," *Communications of the ACM*, 37:21, 23 (Dec.) (stating that "the real purpose behind the proposed digital transmission right is to enable copyright owners to control all digital performances and displays of copyrighted works, without regard to whether they are public or private").

[5]See also McManis, "Emerging Computer Technology," note 64, at pp. 63-73 (disputing White Paper's interpretation of existing law and concluding that it "is highly selective about which uncertainties . . . it would like Congress to address," which "makes critics suspicious about . . . [its] motivation in proposing that the distribution right be amended so as to expressly apply to transmissions").

[6]Cf., *American Geophysical Union v. Texaco Inc.*, 60 F.3d 913 (2d Cir. 1995) (finding that systematic photocopying of an academic journal by a commercial research institute was not fair use where blanket licenses were available from a collection society); *Sega v. Accolade* (9th Cir. 1992) (decompilation of computer programs to extract unprotected ideas was analytical fair use where ideas were not otherwise accessible and expression was not copied). See also *National Basketball Association vs. Motorola*, 1997 U.S. App. Legis. 1527 (2nd Cir. 1997) (holding that real-time transmission of NBA game scores and statistics taken from TV and radio broadcasts in progress to hand-held pagers did not violate state unfair competition law).

[7]See, e.g., McManis, "Emerging Computer Technology," note 64 in text, at pp. 266-69; Jaszi, note 15 in text.

cultural bargain previously depended.[56] Equally uncertain is the role that libraries will play once information providers "can connect directly to the user" via digital transmission over telecommunications networks.[57] Some observers see the changing role of libraries as grounds for allowing publishers virtually unfettered discretion to impose contractual conditions on library access to networked transmissions.[58] Others see the dependence of users everywhere on digital transmissions for the future acquisition of scientific knowledge as grounds for generalizing some of today's library and fair use privileges to all on-line users.[59] The real question, then is how to recreate a "fair use" zone in cyberspace[60] that protects the strong public interest in ensuring that certain uses and certain users, notably the research and educational communities, are not priced out of the market or forced to cut back on the kind of basic research that has heretofore played a crucial role in U.S. economic and technological growth.[61]

THE DRIVE FOR LEGAL PROTECTION OF NONCOPYRIGHTABLE DATABASES

In response to the perceived gap in the worldwide intellectual property system,[62] proposals are being put forward to protect noncopyrightable databases by means of *ad hoc* or *sui generis* intellectual property regimes—that is, by special intellectual property laws that deviate significantly from the classical patent and copyright models[63] that underlie the Paris and Berne conventions of 1883 and 1886, respectively. The impetus for a *sui generis* database law has come from the Commission of the European Communities, whose member countries have adopted, to varying degrees, a policy of commercializing government-generated data.[64] That policy is contrary to the traditional policy of the United States, which has favored full and open access.

Starting in the late 1980s, the Commission of the European Communities began to reevaluate the legal status of databases, especially electronic databases, in the process of formulating an overall strategy for information technologies known as the Information Market Policy Action (IMPACT) program.[65] The Commission found that European database producers had to overcome several comparative disadvantages in order to expand their share of the world market and to catch up with U.S. industry, which dominated the market and was growing at a faster rate than its European counterpart. To overcome these disadvantages, the Commission stressed the need for a single, integrated market, undistorted by differing regulatory approaches, and for higher levels of intellectual property protection, tailored to the needs of potential investors in database production, that might stimulate additional investment in this sector.[66] Another likely premise in the Commission's thinking was that privatizing the government's role in the collection and distribution of data might also generate income streams that could help to offset the shrinking availability of public funds for research and development.

The Commission decided both to harmonize the domestic copyright laws insofar as they applied to compilations of data and to require that the member states adopt *sui generis* intellectual property laws to protect the contents of noncopyrightable electronic databases, a proposal that was subsequently extended to databases in print media as well.[67] In this regard, comparative law revealed that the Nordic countries were already experimenting with short-term, copyright-like protection of noncopyrightable compilations (known as the Nordic catalogue rule), with a view to curbing commercial piracy without extending full copyright protection to borderline literary productions that lacked creative authorship.[68] Accordingly, in 1992, the Commission proposed an innovative Directive to protect such databases, "loosely modeled on the Nordic catalogue rule, [and that] more directly and strongly protects electronic information tools."[69] A greatly amended version of this proposal was adopted by the Council of Ministers and the European Parliament in July 1995,[70] which became the final European Directive on the Legal Protection of Databases of March 11, 1996.[71]

While the precise mesh of the 1996 Directive's two-tiered provisions in administrative and judicial practice remains to be seen, its highly protectionist attributes are unmistakable. Moreover, if all the pending legislative projects are implemented as explained below, similar changes could be introduced into U.S. law and—via a WIPO Database Treaty—into international law as well.

Development of the European Approach to Noncopyrightable Databases

Collections of data, including those relatively unstructured or unprocessed collections of primary interest to scientists, have never fit comfortably within the romantic notion of authorship that once dominated European copyright law, or even within the more pragmatic conceptions of "originality" that pervade modern copyright laws, such as that of the United States. Behind this conceptual resistance there lies a profound concern, often expressed in judicial decisions and the writings of jurists, that facts and ideas constitute building blocks of intellectual discourse that should not (and perhaps cannot constitutionally) be removed from the public domain. In this context, the scientific community's own commitment to the full and unrestricted flow of data represents an important subchapter in a larger discourse that, in this country, at least, is rooted in the First Amendment.[72]

The Commission of the European Communities initially addressed with commendable caution the perceived need for legal incentives to spur investment in electronic database production. The Commission affirmed its preference for a regime based on modified liability principles, that is, one that would deter certain types of socially undesirable conduct without vesting exclusive property rights in data as such (see Box 5.2).[73] Unfortunately, even the Commission's earliest proposals along these lines were flawed by contradictory elements drawn from the exclusive rights model, while the final version became a much less balanced and potentially anticompetitive exclusive property right.[74]

The European Commission's Initial Project

The European Commission's initial approach was premised on the "absence of a harmonized system of unfair competition legislation" to safeguard "the investment of considerable human, technical and financial resources" in the making of databases that "can be copied or accessed at a fraction of the cost needed to design them independently."[75] The logical solution, therefore, was to codify a new type of unfair competition law. Such a *sui generis* law, loosely modeled on existing laws that already protected trade secrets or confidential information, would repress conduct amounting to the "misappropriation" of an electronic database producer's investment without imposing either legal barriers to entry or the social costs of actual or legal secrecy.[76] To this end, the Commission proposed simply to forbid the "unfair extraction" of data from an electronic database for commercial purposes without the second comer's having expended independent effort to collect and verify similar information. The first draft Directive accordingly provided a 10-year period of lead time in which the database maker could recoup his or her investment in a noncopyrightable electronic database while preventing copiers from engaging in for-profit extraction or reutilization of the factual contents, in whole or in substantial part.[77]

The Commission's "unfair extraction" criterion seemed to invite case-by-case judicial distinctions between procompetitive activities, especially independent investment in the generation of a competing electronic database (which was roughly analogous to reverse-engineering by honest means), and market-distorting forms of electronic copying (which were roughly comparable to industrial espionage, commercial bribery, and other types of "parasitic" or free-riding behavior that unfair competition laws interdict). It also may have opened the door to case-by-case judicial evaluation of unauthorized extractions deemed "fair" because they advanced noncommercial educational and scientific pursuits, although neither the draft Directive nor the first Explanatory Memorandum specifically endorsed this proposition. In any event, the drafters further diluted the database maker's new right against "unfair extraction" by engrafting some express user's rights upon it and by adopting explicit measures to safeguard the public interest in free competition.

For example, the drafters apparently envisioned that lawful users of an electronic database could make a limited reuse of its contents even for some commercial purposes, as might occur in value-adding uses. There was also no clear means for database creators to extend the duration of control over the initial compilation by making subsequent changes to it, although the extent to which this omission resulted from a drafting oversight remains unclear. Moreover, price competition was directly encouraged. Second comers could choose between independently compiling their own databases from scratch or invoking a statutory compulsory license against any sole-source provider of data in electronic databases, with a view to competing against that provider while paying reasonable royalties for the use of the data thus extracted.[78] The liability principles loosely embodied in the first draft of the Directive thus created no legal barriers to entry. Arguably, these principles may even have lessened existing economic barriers to entry by empowering would-be competitors to borrow data at reasonable rates when the cost of independently regenerating them appeared too costly or otherwise inefficient as a business strategy.[79]

Absent from this framework were any explicit exceptions favoring educational and scientific users (assuming these were not implicitly "fair" uses under the basic "unfair extraction" criterion of the draft Directive), an omission that the European Parliament singled out for criticism. Although the legislative history does not explain why the drafters rejected this criticism,[80] the reasonable inference from all the evidence is that the Commission believed further exceptions and immunities would unduly weaken the publishers' incentives to invest under a regime that already implemented a procompetitive strategy. If so, the Commission appears to have erred in at least two respects.

First, it seems to have assumed that a competitive market would intrinsically satisfy the needs of the scientific and educational communities, whereas this report shows that basic science has organizational and operational needs that often differ from those a competitive market is geared to meet.[81] Indeed, experience demonstrates that basic science may not be able to pay the market rate even when it is competitively determined. Important research projects consequently may languish for lack of affordable data unless nonmarket mechanisms (such as subsidies) or legal constraints on publishers (such as fair use exceptions and compulsory licenses) close the gap.

Second, the drafters apparently assumed that their concern for the public interest in free competition was still a paramount legislative value in developed market economies. Unfortunately, such protection of the public interest that was implicit in their rudimentary liability framework ultimately gave way to a potent exclusive property right in which public interest safeguards were deemphasized in favor of the protection of private interests.

The European Union's Final Product—The 1996 Directive on the Legal Protection of Databases

Although the European Commission's initial project had undergone transformation by the time that the Amended Proposal was put forward in 1993,[82] its wholesale conversion from a relatively weak liability regime to a strong exclusive property right occurred during the closed proceedings of the European Council of Ministers, which produced the Common Position of July 10, 1995.[83] This version, with minor technical alterations, became the final Directive on Databases, adopted on March 11, 1996, which the European Union member states must now convert into domestic intellectual property laws and regulations.[84]

The Directive as finally adopted may be subdivided into five parts: (1) a list of 60 "recitals" or premises that underlie this legislation; (2) a small group of definitional articles that apply across the board (Articles 1-2); (3) a set of provisions harmonizing the treatment of databases under the member states' domestic copyright laws (Articles 3-6); (4) a set of provisions requiring these same states to provide the new, *sui generis* intellectual property right for noncopyrightable databases (Articles 7-11); and (5) a final group of "common provisions" that apply to both copyright and the *sui generis* laws (Articles 12-16).

Of these, the broad definition of "database" in Article 1(2)[85] constitutes an important feature. Whereas earlier proposals had limited *sui generis* protection to *electronic* databases that were deemed particularly vulnerable to rapid duplication, the broadened definition now includes databases in print form that are accessible to the human eye. In other words, no database circulating within the European Union will escape the regulatory effects of the 1996 Directive, regardless of the medium in which it appears or the nature of its compilers.

As finally enacted, the *sui generis* right conferred on qualifying database makers is no longer couched in terms of "unfair" or even "unauthorized" acts or uses.[86] Rather, the database maker obtains an absolute exclusive "right to prevent extraction and/or reutilization of the whole or of a substantial part, evaluated qualitatively and/or quantitatively, of the contents of that database."[87] This two-pronged exclusive right, which now applies to both electronic and nonelectronic databases,[88] lasts for an initial period of at least 15 years. Any compiler who makes a database available to the public, however, may continually renew the right for additional 15-year terms with every additional investment in the database.[89] This renewal right covers the contents of the entire database, and not just the new matter (as would occur under the derivative work right of copyright law).

The final 1996 Directive does not make *sui generis* protection contingent on the showing of a creative achievement or of a novel contribution to the prior art, the classical bases for justifying legal derogation from free competition in the past. Rather, it requires the database maker to prove that "there has been qualitatively and/or quantitatively a substantial investment in either the obtaining, verification or presentation of the contents" or in "any substantial change resulting from the accumulation of successive additions, deletions or alterations."[90] Because the Directive itself provides no further guidelines for evaluating the requisite level of investment in either case, this threshold will remain uncertain, pending decisions by European courts applying the still to be drafted domestic database laws. Nevertheless, there are no limits to the number of quantitative or qualitative changes that will thus qualify for such extensions, and any publisher who continues to make a substantial investment in updating, improving, or expanding an existing database can look forward to perpetual protection.

Even though the *sui generis* right depends on mere investment rather than on some palpable creative contribution, the scope of protection that the 1996 Directive affords investors in noncopyrightable databases now appears roughly equivalent to that which it elsewhere affords full-fledged authors of copyrightable compilations (and greater than that traditionally afforded to authors of literary works; see Box 5.4).[91] This conclusion follows both from the definitions of the exclu-

BOX 5.4
The 1996 European Directive's Broad Protection for Database Investors

The investor's scope of protection under the hybrid right to prevent extraction appears paradoxically to exceed even that afforded authors of traditional literary and artistic works under the classical copyright paradigm of the Berne Convention[1] in at least three important respects. First, the basic idea-expression dichotomy underlying U.S. copyright law[2] (which the Agreement on Trade-Related Aspects of Intellectual Property Rights (TRIPS) applies universally to all copyrightable works, including such borderline works as computer programs and factual compilations[3]) does not apply to noncopyrightable databases covered by the *sui generis* regime. For this and other reasons explained below, in the universe of data generators, there is no public domain substratum from which either research workers or second comers are progressively entitled to withdraw previously generated data[4] without seeking licenses that may or may not be granted.

On the contrary, every independent generation of data, however mundane or commonplace, will obtain protection if it costs money, and every regeneration or reutilization of the same data in updates, additions, and extensions that cost money will extend that protection without limit as to time.[5] As a consequence, third parties will rarely be able to avoid the expense of regenerating preexisting data—in the way that they can always use previously generated ideas, however much it cost to develop them—unless the originator of the relevant database has abandoned it, or declined to exercise his or her proprietary rights, much as occurs under trademark laws.[6] To be sure, data providers (including, where feasible, members of the scientific community) could decide not to exercise proprietary rights in certain databases, for example, those funded by government agencies; but this would not change the legal situation with respect to scientifically important data located in privately owned databases or in those funded by public agencies, especially foreign agencies, that had opted to commercialize their data.[7]

The absence of any equivalent of the idea-expression doctrine under the new *sui generis* regime means that investors, in effect, obtain proprietary rights in data

---

[1]See, e.g., Reichman, "Collapse of the Patent-Copyright Dichotomy," note 8 in text, at pp. 485-86, 492-96; see also Sam Ricketson (1987), *The Berne Convention for the Protection of Literary and Artistic Works: 1886-1986*, Queen Mary College, at pp. 231-32.

[2]See 17 U.S.C. §102(b) (1994); text accompanying note 16 in text.

[3]See TRIPS Agreement, note 111 in text, articles 9(2), 10(1); Reichman, "Know-How Gap in TRIPS," note 120 in text, at pp. 775-84.

[4]Cf. Jaszi, note 15 in text, at p. 596; Litman, "Public Domain," note 9 in text, at p. 967.

[5]See notes 89-90 in text.

[6]See Lanham Trademark Act, 15 U.S.C. §1052 (1994 ed.). However, there is an infinite array of trademarks, and the use of marks to distinguish quality producers inherently promotes competition without creating legal barriers to entry. See, e.g., Stephen Ladas (1975), *Patents, Trademarks, and Related Rights*, Harvard University Press; William M. Landes and Richard A Posner (1987), "Trademark Law: An Economic Perspective," *J. L. Econ.*, 30:265.

[7]Executive Office of the President, Office of Management and Budget, "Implementing the Information Dissemination Provisions of the Paperwork Reduction Act of 1995," Memorandum by Alice M. Rivlin, September 22.

as such, a type of ownership that the copyright paradigm expressly precludes. The drafters of the *sui generis* right play this down by insisting that third parties always remain free to generate their own databases. But this opportunity exists only for data that are legally available from public sources and whose cost of independent regeneration is not prohibitively high in relation to the gains expected from the exercise, if any. As for proprietary data not legally available to second comers to exploit, there is no opportunity to avoid the originator's exclusive rights to prevent extraction or reuse of existing data. In such cases, the investor's exclusive rights necessarily vest in the data as such.

A deeper point is that, regardless of whether it is possible in theory to regenerate the data from publicly available sources, investors in database production can always deny third parties the right to use preexisting data in value-added applications,[8] even when the third parties are willing to succumb to royalty-bearing licenses;[9] and there is no escaping such licenses unless the database publisher either declines to exercise his or her rights, or engages in an abusive exercise of market power. In other words, except when the new proprietary rights are abandoned or misused, the concept of incremental or "cumulative and sequential innovation," which is central to the development of modern technological paradigms,[10] has been banished from the universe of database production, despite the economic waste and inefficiency inherent in such policies.

A second, and closely related way in which the database investor's scope of protection under the 1996 European Directive exceeds that of authors under the classical copyright paradigm is to be seen in the treatment of derivative works. Under copyright laws, the scope of an author's exclusive right to make a derivative work extends only to the original, expressive matter added to the underlying work. It cannot protect either ideas or preexisting expressive matter, including any matter that has entered the public domain.[11] But the 1996 Directive recognizes no such legal distinctions and, as just explained, it harbors no working conception of a public domain whatsoever. Unless local European courts applying the domestic laws that implement the Directive take pains to limit this omission, the upshot will be that each new extension of the database maker's exclusive rights by dint of his or her "substantial investment" in updates, additions, and revisions[12] will, in effect, requalify that investor for protection of the database as a whole, for additional 15-year periods, and not just for the revised or added matter—the "derivative work"—as would occur under the copyright laws. This, in turn, reinforces the monopolistic

---

[8]See, e.g., Samuelson, "Missing Foundations," note 34 in text.

[9]However, a refusal to license, coupled with a dominant position in the marketplace, could trigger an antitrust violation or a related charge of abuse of intellectual property rights. See, e.g., E.C. Directive on Databases, note 52 in text, article 16(3); Hunsuker, note 32.

[10]See, e.g., Richard R. Nelson and Sidney G. Winter (1982), *An Evolutionary Theory of Economic Change*, pp. 255-62; Richard R. Nelson (1994), "Intellectual Property Protection for Cumulative Systems Technology," *Colum. L. Rev.*, 94: 2674, 1676; see also Robert P. Merges and Richard R. Nelson (1990), "On the Complex Economics of Patent Scope," *Colum. L. Rev.*, 90:839, 881.

[11]See, e.g., 17 U.S.C. §§101, 102,103, 106, 501 (1994).

[12]See notes 89-90 in text.

(continues)

> **BOX 5.4 Continued**
>
> effects inherent in the originator's ability to deny third parties the right to build incrementally and sequentially upon preexisting scientific and technical knowledge.
>   A third way in which the scope of protection for investors in database production exceeds that afforded authors of copyrightable literary and artistic works results from the much narrower range of public interest exceptions applicable to investors.[13] In effect, the sole important exception available to all users of noncopyrightable electronic databases under the Directive is the right to extract or reutilize "insubstantial parts of the database."[14] Even this exception applies only to "lawful users" of the database (i.e., presumably subscribers to an on-line service or purchasers of a CD-ROM), which suggests that in most cases, the exception merely validates acts incidental to obtaining the value for which one has paid. Moreover, a lawful user of a noncopyrightable database cannot extract or reuse even insubstantial parts of its contents in "repeated and systematic" ways that "conflict with a normal exploitation of that database or . . . unreasonably prejudice the legitimate interests of the maker."[15] Arguably, this could preclude most value-added uses of an insubstantial part of the database, regardless of the commercial or noncommercial purpose of the users.
>
> ---
>
> [13]For the general range of public-interest exceptions under copyright laws, see text accompanying notes 16-20 in text.
> [14]See E.C. Directive on Databases, note 52 in text, article 8(1). However, member states may allow "extraction for private purposes of the contents of a *non-electronic* database." Id., article 9(a).
> [15]E.C. Directive on Databases, note 52 in text, article 7(5); see also id., article 8(2) (forbidding any acts by lawful users that "conflict with normal exploitation" or "unreasonably prejudice the legitimate interests" of its maker).

sive rights set out in the Directive itself and from the Council of Ministers' closed-door decision to delete from its 1995 Common Position the initial draft Directive's compulsory license requirement facing sole-source providers.[92]

As defined in article 7(2) of the 1996 Directive, the investor's *sui generis* extraction right covers even temporary transfers of data to on-line receivers, much like the author's broadened rights to prevent reproduction in copyright law under article 5(a).[93] The investor's reutilization right covers on-line use or transmissions of data, including those in value-added or derivative formats, much like the author's broadened "communications" rights under article 5 (b), (d), (e).[94] In this and other respects, including the omission of any requirement for a compulsory license against sole-source providers, the drafters of the 1996 Directive have integrated the *sui generis* regime into the broader regulatory framework for national and international information infrastructures that the European Union and U.S. intellectual property authorities are now jointly promoting.[95]

Potential Effects of an Exclusive Property Rights Approach

Unlike earlier versions, the final 1996 Directive does give European Union member states the option of allowing authorized extraction of a substantial part from a noncopyrightable database "for the purposes of illustration for teaching or scientific research, as long as the source is indicated and to the extent justified by the noncommercial purpose to be achieved."[96] This exception is available only to a "lawful user" and only for the purpose of "extraction" but not for that of reutilization, and it will exist only in those member states that opt to enact it.[97]

If a member state enacts this provision, the scientific or educational user's exempted extraction must satisfy both the noncommercial purpose test and the general obligation "not [to] perform acts which conflict with normal exploitation or [that] unreasonably prejudice [the maker's] legitimate interests."[98] Because the normal use of a scientific database in an academic institution typically is to serve as a research and educational tool, this exemption could merely permit illustration of conclusions reached, but not uses for other scientific or educational purposes, such as browsing or even extractions from and use of the collected data for the purpose of reaching the conclusions that one may then freely "illustrate."[99]

Of course, local legislators might manufacture loopholes through which to widen this exception,[100] and database publishers might refrain from imposing harsh or oppressive terms and conditions that unduly impinge upon scientific and educational uses. The fact remains, however, that nothing in the Directive as finally enacted requires such accommodations. Its *sui generis* provisions contain no real equivalents of the private use, fair use, and related exceptions that traditional copyright laws afford scientific and educational users of core literary and artistic works. Moreover, database publishers who acquired market power through restricted on-line transmissions reportedly have recently imposed questionable contractual conditions on libraries and academic subscribers.[101]

It follows that under the 1996 Directive, the most borderline and, in the sense that they are basic building blocks of knowledge, questionable of all objects to receive intellectual property protection—compilations of data and facts, scientific or otherwise—paradoxically obtain the strongest scope of protection available from any intellectual property regime except, perhaps, for the classical patent paradigm itself.[102] Nor are the breadth of protection and the monopolistic power it tends to breed likely to be offset by greater competition in the market for electronic databases, especially now that the 1996 Directive as finally adopted no longer contains the compulsory license requirement that had initially been devised for this purpose.

Formally, of course, third parties still remain free to compile a database exactly like one already in commerce, because independent generation of the relevant data at one's own time and expense is always permitted. In practice, this option ignores the economic realities of the database industry, in which start-up costs can be relatively high, the prospects for market sharing have seldom been realized, many valuable data sets are unavailable from public sources, and the existence of one complex database seems empirically to constitute a *de facto* barrier to entry that is seldom overcome. Moreover, as discussed in previous chapters, many databases in the natural sciences contain unique, nonreproducible observations that are by definition available only from a sole source. This lack of effective competition, with its inherent possibilities for discouraging add-on products and for engaging in abuses of market power, was downplayed by the European Council of Ministers in 1995, even though it had been uppermost in the minds of the European Commission's own drafters a short while earlier. Article 16 of the final Directive thus merely calls for 3-year reviews to determine whether existing antitrust laws prove inadequate to deal with the "abuse of a dominant position or other interference with free competition," in which case proposals for "non-voluntary licensing" may once again be considered.[103]

The fear of market failure and of chronic underprotection that initially motivated the quest for a *sui generis* regime to protect electronic databases has thus given way to the creation of "mini-monopolies over information"[104] and to an underlying logic that is inconsistent with the public interest in the full and open flow of scientific data. The original goal of providing some incentives to augment the publishers' investment in compiling electronic databases has generated a set of norms that could render many scientific and technological undertakings prohibitively expensive. As explained below, the short-term social benefits of this so-called "extraction right" may thus conceal the long-term social costs of diminished research and development capabilities at scientific and educational institutes, including public and semipublic institutions that are already indirectly subsidizing private research and development.[105]

Overlapping U.S. and European Union International Models

When the European Commission began its deliberations concerning database protection, the climate in which intellectual property policy discussions at both the national and the international level took place differed from that prevailing today. The fate of the Uruguay Round of multilateral trade negotiations and its intellectual property component, the Agreement on Trade-Related Aspects of Intellectual Property Rights (TRIPS), remained uncertain. The U.S. intellectual property authorities had not yet begun to survey the issues posed by widespread transmission of digitized information over telecommunications networks. The Supreme Court had just denied copyright protection to telephone directories in *Feist* and had recently invalidated state protection of subpatentable industrial designs.[106] These decisions proclaimed renewed faith in a 19th-century vision of the competitive ethos without recognizing the unresolved problems of gaining returns from investments in subpatentable information goods under 21st-century conditions.[107] It also seems noteworthy that a few years earlier, the chairman of the House Subcommittee on Intellectual Property had set very high standards that would have to be met before Congress would consider proposals for additional forms of *sui generis* intellectual property protection that deviated from the classical patent and copyright paradigms.[108]

Against this background, the European Commission's early drafts for a Directive on databases adopted a defensive posture with respect to foreign publishers, which proposed a strict criterion of material reciprocity. Databases made in countries that did not enact *sui generis* legislation akin to that envisioned by the Directive would consequently remain vulnerable to wholesale copying within the European Union itself.[109] This decision to discriminate against foreign nationals operating in nonharmonizing states was modeled on the earlier and equally controversial decision by the United States to impose a material reciprocity clause under the Semiconductor Chip Protection Act of 1984.[110] Although both decisions rested on dubious legal grounds even before the TRIPS Agreement was adopted, and even though that agreement rejects this approach, at least in spirit,[111] a version of the reciprocity provision nonetheless entered the European Commission's final Directive on the Legal Protection of Databases, as adopted in 1996.[112]

A Coordinated High-Protectionist Strategy

By 1995, however, when the European Union's Council of Ministers met to adopt its Common Position on the pending database Directive, the climate surrounding worldwide intellectual property policymaking had profoundly changed. Universal intellectual property standards embodied in the TRIPS Agreement had become enforceable within the framework of the World Trade Organization,[113] largely as the result of sustained pressures by a coalition of powerful manufacturing associations in Europe, the United States, and Japan.[114] The success of this venture presages further alignments of interests by U.S. and European Union officials with a view to forging a common, strongly protectionist strategy for intellectual goods in the post-TRIPS environment.[115]

Included within this strategy was a packet of complementary proposals for amending or expanding the Berne Convention. Known as the "Digital Agenda," these proposals were considered in a December 1996 Diplomatic Conference hosted by WIPO.[116] Some of the proposals, which the European Union's own intellectual property authorities placed on the agenda for that conference, would have conformed international copyright law to the regulatory framework for a global information infrastructure that the U.S. Information Infrastructure Task Force's (IITF) White Paper on the national information infrastructure recently endorsed.[117] The White Paper took the view that on-line providers are, or should be, strictly liable for digital transmissions of copyrighted works, even if this obliges providers to serve as "copyright police" without regard to their ability to perform such functions.[118] It also proposed a battery of measures that would prohibit the decoding of encrypted transmissions (or the tampering with other electronic safeguards) as copyright infringement and that would forbid altering "copyright management information, including the terms and conditions for access to on-line transmissions."[119]

Despite the innocuous appearance of these and related proposals, they are broadly drafted and may result in indirectly overruling numerous judicial precedents, including some that permit reverse-engineering of the noncopyrightable components of computer programs.[120] These proposals might also help to immunize copyright owners from claims of misuse for imposing harsh or oppressive conditions on users in the form of nonnegotiable electronic contracts.[121] The single most troubling aspect about the White Paper (and the legislative proposals it has spawned) is, as so many qualified observers have concluded, that it favors "reducing the application and scope of the fair use doctrine in cyberspace."[122]

Another proposal, the WIPO Draft Database Treaty,[123] called for worldwide protection of noncopyrightable databases under *sui generis* intellectual property regimes. A Committee of Experts prepared the WIPO Draft Database Treaty, which draws from both the European Union's model and from draft U.S. legislation. It was placed on the December 1996 Diplomatic Conference agenda at the behest of the U.S. Patent and Trademark Office (PTO).[124] The Diplomatic Conference was thus asked to convert the WIPO Draft Database Treaty into international law, even though the United States lacked any corresponding domestic regime as of the time of its writing.[125] In addition, there has been no empirical test of the controversial final 1996 European Directive in actual practice,[126] and no preliminary reports or studies evaluating even the economic justification for such measures have been issued by WIPO or by any other reputable international institution.[127] The Diplomatic Conference postponed action on this proposal and charged WIPO to set a timetable for further deliberations.

Against this background, the changes to the European Commission's Database Directive made in the Council of Ministers' Common Position of 1995, including deletion of the compulsory license provision and other measures that strengthened the exclusive rights apparatus,[128] reflect the coordinated strategies that the European Commission and the U.S. intellectual property authorities are now jointly pursuing. If implemented as proposed, these strategies could gradually extend international norms concerning the legal protection of databases from the Berne Convention (or related instruments) to the TRIPS Agreement, which empowers the Council for TRIPS to "undertake reviews in the light of any relevant new developments which might warrant modification or amendment of this Agreement."[129] This, in turn, could eventually obviate the long-term effects of the reciprocity clause in the 1996 European Directive by replacing it with a set of harmonized norms binding on all World Trade Organization member states, like those already adopted for semiconductor chip designs in articles 35-38 of the TRIPS Agreement.[130]

H.R. 3531—The Proposed U.S. Database Investment and Intellectual Property Antipiracy Act

The bill proposing a U.S. law to protect noncopyrightable databases, which was introduced to the House of Representatives at the behest of the PTO,[131] articulated a still more protectionist strategy than that of the European 1996 Directive. Under H.R. 3531, as under the final 1996 Directive, a compiler would qualify for exclusive rights to prevent extractions and reuses of the whole or substantial parts of any database by dint of his or her having made substantial investments in the collection, assembly, verification, organization, or presentation of its contents.[132] These exclusive rights would attach automatically upon the expenditure of resources, and if the owner continued to invest in updating or otherwise maintaining the database in question, its 25-year initial term of protection could be renewed continually, without limit,[133] in all the contents of that database. This provision thus ignores the constitutional enabling clause, which requires intellectual property rights to be limited in time.[134]

Furthermore, when one scrutinizes the details of the pending U.S. proposal, one finds that its definition of "database" is much broader than that of the 1996 Directive. It contemplates, for example, that noncopyrightable components of computer programs could qualify for protection as databases, and it provides no apparent criterion for excluding even facts or data compiled for scientific and historical works.[135] Moreover, the database maker's exclusive rights to extract, use, or reuse all or a substantial part of the contents are reinforced by allowing database makers to control any use that "adversely affects the actual or potential market for that database" in addition to uses that otherwise "conflict with the database owner's normal exploitation."[136] This specification, which is not found in the 1996 Directive, has the potential for impeding virtually any judge-made exceptions analogous to "fair use" under copyright laws, because any such exception would almost certainly affect the "potential market" for any given database.[137] At the same time, the database owner's potentially perpetual "derivative work" right (flowing from continuing updates), which is subject to no public domain exceptions whatsoever, becomes even easier to obtain than under the 1996 Directive, because the U.S. bill would condition the renewal right merely upon "any change of commercial significance" to the database contents and not solely on additional "substantial investments."[138]

The pending bill then expands the U.S. database owner's scope of protection well beyond that of the 1996 Directive's *sui generis* regime by introducing an array of measures that, when read together, could produce formidable anticompetitive effects. For example, the 1996 Directive's principal concession to users—the exception for extraction of insubstantial parts[139]— is ostensibly broadened in H.R. 3531 to permit uses or reuses of insubstantial parts,[140] but it is then significantly narrowed in at least two ways. First, there is a new provision that not only forbids "repeated or systematic use or reuse of insubstantial parts" (like the comparable provision of the European Directive),[141] but also expressly forbids extraction or uses even of insubstantial parts "that cumulatively conflict . . . with . . . normal exploitation . . . or adversely affect . . . the actual or potential market."[142] This latter clause acquires further strength by means of still other provisions that seem to outlaw extraction or reuse of even insubstantial parts of a protected database in any product or service that directly or indirectly competes with the database from which it was extracted in any market, however distant.[143] Also forbidden are extraction, use, or reuse of even insubstantial parts "by or for multiple persons within an organization or entity in lieu of . . . authorized additional use or reuse . . . by license, purchase, or otherwise."[144]

Given such restrictions, one is hard-pressed to imagine unauthorized uses of an insubstantial component that the drafters of the U.S. bill would deem legitimate. To forestall even this remote possibility, the bill allows publishers contractually to override even the formal right of lawful users to extract or use insubstantial parts, in contrast with the express nullification of similar contractual provisions in the 1996 Directive.[145] One knowledgeable source reports that some U.S. database publishers, opposed to this constraint in the 1996 Directive, expressed an intent to exercise permissible contractual overrides in practice.[146] A similar intention seems manifest in the clause allowing publishers to impose separate licenses for networked use of a database within organizations, including nonprofit academic and scientific institutions, which can be construed as covering the extraction, use, or reuse even of insubstantial parts.[147]

Taken together, these and other provisions of the proposed H.R. 3531 reinforce the single most disturbing aspect of the 1996 European Directive, namely, that it precludes formation of an evolving public domain from which third parties can freely draw.[148] To this end, the bill expressly confines permissible acts of "independent creation" to data or materials not found in a database subject to the proposed *sui generis* regime.[149] This restriction applies regardless of whether the unauthorized extraction or use is made for purposes of noncommercial scientific endeavor or for commercially important value-added products that build incrementally on existing compilations of data. Every unauthorized use or reuse of existing data thus potentially violates the database owner's unbounded derivative work right. Furthermore, the existence of this potential violation is determined without regard to the substantiality of the second comer's own expenditure of effort or resources, to the similarity or differences of the latter's product or service, or to the public-good aspects of the activities undertaken.[150]

The monopoly conferred on database owners under the pending U.S. legislation is then perfected by recognizing no public interest exceptions whatsoever. Even the weak exception for extraction (but not reuse) "for the purposes of illustration for teaching or scientific research" that the 1996 Directive allowed European Union member states to enact[151] is omitted from both H.R. 3531 and the U.S. submission to WIPO.[152]

The sole concession to science and education in H.R. 3531 is a provision not found in the European Directive that expressly denies coverage to "a database made by a [federal] government entity."[153] Because most databases of primary importance to science are funded by federal government agencies, this provision appears to recognize that such databases merit different treatment from those normally covered by the proposed *sui generis* regime. The message is rendered ambiguous, however, because it does not expressly apply to databases funded by government and also because of language in the same provision to the effect that "any database otherwise subject to this Act . . . is not excluded herefrom because its contents have been obtained from a governmental entity."[154] This language appears to allow private firms that invest in data obtained from federal government sources to qualify for protection. However, it also can be read as implicitly inviting federal governmental agencies to derogate from the traditional U.S. position, reiterated in a recent Office of Management and Budget (OMB) directive, which limits such agencies to the recovery of dissemination costs from commercial applications of government-funded data by the private sector.[155] If government agencies were to move beyond the cost-recovery threshold, the continued ability of scientists to access such data on favorable terms, which current policy seeks to guarantee, would then be called into question.

Disregarding the status of databases made by governmental entities, H.R. 3531 would render virtually any act of "collecting, assembling, or compiling . . . data . . . from . . . a database subject to this Act" a prohibited or infringing act. The perpetrator could never justify such acts as incidental to other acts of independent creation, or as incidental to recognized public interest exceptions, or even as a legitimate means of building on preexisting data sets.[156] Nor does H.R. 3531 express any concern that application of its exclusive rights might lead to abuse of a dominant position or to other anticompetitive acts that might require "nonvoluntary licensing" at some point in the future.[157]

Ancillary provisions of H.R. 3531 also embody some of the current administration's most controversial proposals concerning the regulation of national and global information infrastructures. For example, one provision, following a proposal from the IITF White Paper, would outlaw making or distributing any technical device (or performing any technical service) the primary purpose of which is to circumvent self-help technological security measures that publishers rely on to help protect the contents of their databases.[158] Another provision, also inspired by the White Paper, would forbid tampering with database management information attached to copies of database contents or otherwise distributing copies in a form that bears false information about ownership or other aspects of managing the relevant proprietary rights.[159]

Self-help measures, such as encryption for networked transmissions, often serve valid commercial purposes, and they may be indispensable for the protection of privacy.[160] However, such measures also may unduly reinforce the publisher's power to impose harsh contractual terms in two-party deals,[161] a prospect that H.R. 3531 completely ignores. There are concerns as well that publishers will use these provisions to fend off legitimate public interest challenges to the scope of protection obtained under the proposed *sui generis* regime. If, for example, decrypting a coded transmission is necessary to extract part of a database for noncommercial scientific purposes, and that act of decryption itself constitutes a tort,[162] researchers are unlikely to explore the potential availability of judge-made public interest exceptions to the exclusive rights conferred by the new regime.[163]

In sum, by providing a longer period of protection, more powerful exclusive rights, no public interest exceptions or privileges, harsh criminal penalties,[164] and ancillary rules reinforcing self-help policing of on-line transmissions, the proposed U.S. law H.R. 3531 would grant database owners greater monopoly power than that emanating from the European Union's 1996 Directive. In so doing, the drafters of H.R. 3531 take no notice of the important role that affordable, unrestricted flows of data have traditionally played in supporting U.S. research and education, or in other sectors vital to economic development. The proposed regime thus risks triggering a chain of unintended consequences that could ultimately compromise both the foundations of basic science and the technological superiority of the national innovation system.[165]

The Rush to Legislate

In mid-1996, the PTO (which now speaks for the copyright office in international affairs) presented proposals similar to those set out in the IITF White Paper before WIPO, with a view to amending the Berne Convention and to adopting new instruments related thereto at the December 1996 WIPO Conference.[166] However, the Diplomatic Conference rejected many of these proposals and adopted a more socially balanced approach to the protection of copyrighted works digitally transmitted over telecommunications networks. Whether WIPO member states—including the United States—will enact this more balanced approach into their domestic laws remains to be seen. In May 1996, the U.S. negotiators at WIPO also presented the draft of an international treaty to protect noncopyrightable databases,[167] and on August 30, 1996, the chairman of the Committee of Experts published a WIPO Draft Database Treaty. The conference postponed immediate action on this treaty, but at a meeting of the WIPO Governing Body in March 1997, a new timetable was established for further work.

This unnecessarily fast pace has so far allowed little time for public hearings or for national science groups and other interested parties to organize, analyze the proposals, and contribute to their shaping.[168] If successful, it would convert the IITF White Paper's "reform proposals" for a *sui generis* law to protect noncopyrightable compilations of data into international minimum standards of intellectual property protection binding on all signatories to the Berne Convention.[169] This "whiplash effect" would then oblige the United States to implement these same standards in its domestic laws, as a matter of international law, even if

Congress had not already adopted similar legislation in the interim.[170] The haste with which both the U.S. and the European Union authorities have moved to implement these measures at the international level[171] thus raises still further questions about the extent to which the public interest has been compromised.

The case for moving so far and so fast rests largely on the supposed difficulties of enforcing territorially grounded intellectual property rights in cyberspace[172] and on the fear of "detaching information from the physical plane, where property law of all sorts has always found definition."[173] From a legal perspective, these developments raise daunting problems of conflicts of law, a field that has never found it easy to accommodate intangible property.[174] Yet, it will not do to exaggerate these difficulties while ignoring the harmonizing effects of the TRIPS Agreement, which requires all countries that belong to the World Trade Organization to adopt *both* the universal minimum standards of the Berne Convention (whether or not they adhere to that convention) *and* the additional standards concerning computer programs, compilations, and related subject matter set out in the TRIPS Agreement itself.[175] Regardless of whose law applies, in other words, digitally transmitted information goods will eventually become subject to the same international minimum standards of protection in all developed countries and in most developing countries as matters stand.[176]

To be sure, these standards harbor "gray areas" that are open to different interpretations, notably with respect to the scope of copyright protection afforded borderline works, such as computer programs and those databases that otherwise meet the domestic criteria of eligibility.[177] But the developed countries have only just begun to grapple with these issues, and there is no basis for an empirically grounded consensus even with regard to computer programs or industrial designs,[178] let alone databases and other electronic information tools. As previously demonstrated, moreover, a solid body of scholarly opinion holds that "a combination of technological restrictions (such as encryption), contractual arrangements and criminal sanctions (for unauthorized decryption)" constitutes overprotection that raises a far greater risk than the risk posed by underprotection.[179] Taking the IITF White Paper's controversial proposals to a premature diplomatic conference without allowing time for adversely affected users, especially the scientific and educational communities, to participate in the discussions thus lacked justification in terms of sound public policy.

CHARTING A WELL-CONSIDERED COURSE IN THE NEW ERA

Putting Science into the Picture

At the time of this writing, neither the scientific nor the educational community has played any part in the relevant deliberations concerning the legal protection of databases, and they have not been consulted on any official basis. If matters proceed without adequate input from researchers and educators, Congress could enact the proposed *sui generis* database regime, despite the risk that "it would allow a limited group of database creators to control the dissemination of information" and that the "resulting restrictions on the transfer of knowledge would be detrimental to society, as information lies at the core of social advancement."[180]

In most cases the proposed *sui generis* regime will simply engraft a strong legal monopoly onto the preexisting natural monopolies that are typical of the database industry. As noted in Chapter 4, it is the social costs of these double-barreled monopolies to the public at large that must actually be taken into account, along with their overall impact on a scientific community whose leading role in world technological production is linked in still unexplored ways to the traditional funding of scientific data by government.

Clearly, U.S. policy makers should not incur such risks without evaluating in advance the possible repercussions that *sui generis* database laws might have on the nation's scientific and technological capabilities and future progress, and without taking measures to alleviate them before embarking on such an uncharted course. By the same token, the scientific community has a vital stake in the formulation of new database laws, in order to ensure that legal incentives to stimulate investment in the production and distribution of data do not end by impeding the full and open flow of those same data to basic science.

The scientific community can ill afford to remain indifferent to these proposals for database protection, if only because its whole *modus operandi* has been based on the principle of full and open exchange of data. This principle is indirectly undermined by the pending proposals concerning legal regulation of the national information infrastructure and directly threatened by the drive to institute *sui generis* intellectual property rights in the contents of electronic and other databases. As regards the database laws in particular, the foregoing analysis suggests that science and education have two paramount concerns that need to be pursued in the course of future legislative deliberations:

- *Sui generis* laws to protect databases should, on the whole, reflect a proper balance between public and private interests, including the public interest in free competition, that is, between public goods and private intellectual property.
- Such laws should contain measures specifically designed to preserve and promote the scientific and educational enterprise, including the need to facilitate and encourage the establishment and maintenance of databases essential to the work of science.

Reconciling the Needs of Science with Those of a Competitive Market

The advent of new proprietary rights where none previously existed will influence the collection and distribution policies of all data providers, including government-funded providers and other sources that currently follow nonprofit pricing policies. As funding sources shrink and foreign governments operating under the European Union's 1996 Directive shift to profit-oriented policies, more and more data of interest to science will be covered by proprietary rights, and fewer data will be made available to science on a cost-of-dissemination basis. The tensions already reflected in the recent OMB circular[181] will become more generalized, even though different disciplines will experience different degrees of hardship.

The experience with Landsat is indicative of transnational problems likely to arise when states adopting different policies invoke their sovereign rights to buttress their respective positions. For example, some of the international ground stations that receive Landsat data reportedly object to the traditional U.S. policy of making data available at the cost of fulfilling a request. They want to continue to charge whatever the market will bear, and they are pressing the United States to change its policy and laws. If the cost-recovery approach is not extended to ground station agreements in other countries, this would leave academics and other nongovernmental users to pay prices that they simply cannot afford.

As discussed in Chapter 4, the adverse effects of Landsat commercialization on the scientific community were easy to document, although the value of lost research opportunities remains hard to quantify in terms of objective social costs. In other cases, however, it will prove harder to show the effects on science, especially if a commercialized database has many private downstream users who are better able to afford the rates, and there is no powerful upstream user community—akin to the global change research users of Landsat data—capable of voicing its distress in terms that cannot be ignored. In such cases, the high cost of data may simply inhibit project formulation when there is no realistic possibility of funding that cost. Yet, because academic scientists are relatively few in number and not typically a presence in day-to-day decision making at the policy level, their lost research opportunities may simply go unreported and unrecorded.[182] Although such lost research opportunities are difficult to predict or quantify accurately, the areas likely to be most adversely affected include data-intensive research in the observational sciences that rely on unique, multiple, or continuously updated data sources.

Of course, the law of diminishing returns also applies, and commercial providers may find that no one will access their files if they charge too much. Before this point is reached, however, the more likely result—as suggested by the Landsat example—is that the provider may determine that the price/volume point on the demand curve at which the service is expected to be viable can be afforded by only a few well-financed scientists. This does not provide general access for those unable to pay at that level, including both students and scholars with limited grant funds, not to mention scientists and other potential users in poorer communities.

More generally, such an approach ignores the contribution of basic science to the ability of U.S. firms to predominate in markets for technology and information goods. Despite a general consensus on the need for sustained levels of investment in research and development, the proposed database laws could change the status quo—without anyone's wanting it to happen—by elevating the price of the one raw material to which U.S. researchers have always had ready access. If less available scientific information were to translate to fewer applications of economic importance, the end result would be a loss of U.S. technological competitiveness in an integrated world market.[183]

Preserving the Public-Good Aspects of Science—What Is Needed?

The negative prospects outlined above do not mean that the interests of research and education are best served by the absence of legal protection for the contents of databases. As Chapter 4 suggests, a socially balanced, procompetitive database regime might indirectly help science to contain costs by bringing market forces to bear on some of the pressure points. It would provide a greater stimulus to third-party investors who might compete with sole-source data generators or distributors (when the market segment in question can feasibly support multiple providers), or who might adapt sole-source data sets to applications of particular interest to science. While this stimulus might not change the overall market structure or significantly reduce the formation of natural monopolies, in the short term at least it could help to trigger countervailing tendencies and thus lead to lower prices and fewer restrictions on access, particularly if novel, value-added products become of greater importance to science over time.

Conversely, if a socially imbalanced, overly protective database law converts existing impediments into insuperable legal barriers to entry, the adverse effects on science—absent offsetting legal safeguards—would soon make themselves felt (see Box 5.5). In this context, the scientific and educational communities—like value-adding users and second comers in general[184]—would arguably fare better either under a simple unfair competition law that prohibits wholesale copying or under a *sui generis* regime built on more refined liability principles than under any regime based on exclusive property rights.

A liability model creates no legal barriers to entry in its own right, nor need it significantly strengthen the sole-source data provider's market power. A liability regime also can eliminate the "refusal-to-deal" problem, by addressing the serious concerns of those who fear the power of sole-source data providers to restrict access to data on a variety of grounds. When an automatic license is built into a modern liability regime, it tends inherently to solve the problem of abuse without recourse to antitrust law. For these reasons, the European Commission's initial preference for a liability regime, rather than an exclusive property right,[185] merits careful consideration by the U.S. scientific community as a possible response to the overall challenge posed by the drive for *sui generis* database laws.

> **BOX 5.5**
> **Possible Consequences of an Overly Protectionist Database Regime**
>
> Under an exclusive property rights model, a database owner's absolute monopoly could disincline him or her to allow scientists access to certain files, especially if the owner feared that the uses in question could lead to value-added products that diminished his or her market power.[1] Providers and distributors would also be likely to charge higher prices for all uses, to demand payment for certain uses that were previously free, and to resist pressures for price discrimination favoring scientific users.
>
> As matters stand, the electronic publishers' growing capacity to charge for each and every use of on-line data (or at least for every documented access to the database) and to track and monitor every user potentially liable for these charges means that it becomes increasingly capable of serving "as its own collection society, subject to no consent decrees, no membership controls and no external regulation."[2] In this milieu, even blanket licenses can be set unrealistically high for large-scale nonprofit users, such as libraries, universities, and research institutions, and the net impact of the licensing fees will further depend on other contractual conditions that accompany the licenses. Even when a blanket license fee is relatively low, for example, if the institution is obliged to purchase many licenses for different researchers or groups of researchers, the total cost may still become prohibitive. The existing tendencies of some publishers to approach academic and scientific users one by one and to impose harsh or oppressive terms[3] could only be strengthened by the enactment of a new and powerful intellectual property right covering the contents of electronic databases as such.
>
> ---
>
> [1] Here one would expect further tensions stemming from the scientific community's own efforts to internalize electronic publication of research results at the expense of both commercial publishers and professional societies. Cf. Paul Ginsparg, note 35 in text.
> [2] Reichman, *Electronic Information Tools*, note 15 in text, at p. 464.
> [3] See note 3 in text (OTA report); Reichman, "Electronic Information Tools," note 15 in text, at p. 464 (noting that licensing agreements "may consequently require libraries to waive privileges" under new or existing laws, including copyright law, "and to limit users' access to [the protected] matter beyond what their own understanding of the . . . [laws] would require. Aggressive licensing of electronic information tools could thus distort the public service mission of libraries by making them involuntary collection agents for publishers"). Cf. also *American Geophysical Union v. Texaco, Inc.*, 60 F. 3d 913 (2d Cir. 1995) (oil company's unauthorized copying of articles from its own library's technical journals for archival use by its own research scientists was not fair use under copyright statute where blanket licensing schemes were available); Patterson and Lindberg, note 21 in text, at pp. 159, 181-90.

With or without the more procompetitive conceptual framework of a liability model, a socially balanced database law should preserve and promote the public-good aspects of science and education. This goal requires careful crafting of its technical legal machinery, as well as the inclusion of safeguards that address the specific needs of the scientific and educational communities.

Ensuring Legal Safeguards for Access

Exclusive control over data, like exclusive control over ideas, raises serious concerns, including First Amendment concerns, that are particularly germane to open scientific inquiry.[186] While meeting these concerns does not necessarily imply that data should become available without charge or proprietary interests, it does mean the following:

- The law itself should define the parameters of an evolving public domain from which investigators can freely extract and use data for certain purposes.
- The law must also guarantee scientific and educational users access to that domain on reasonable terms and conditions.
- The definition of a protectible database should be narrowed so as to exclude ideas and contents of scientific theories.
- Database owners should never possess the right to preclude access to otherwise publicly available data when sought for purposes of basic scientific research.

The terms of access would then depend in part on the size and scope of any free use and fair use zones built into a proper *sui generis* law for the benefit of scientific and educational users.[187]

Publishers are likely to oppose such exceptions because they represent a de facto subsidy to educational and scientific users, which in an on-line environment can no longer be hidden behind the ancillary need to overcome transaction costs.[188] Nevertheless, the case for maintaining such exceptions is even stronger with regard to the contents of databases than to other objects of intellectual property protection. As in other cases, publishers require state intervention in the marketplace to enforce the fictitious portable fences on which the protection of intangible literary productions depends. In this case, however, the objects of protection—data—are functionally determined elements or particles of knowledge that fall well below the "grain size" threshold of existing intellectual property laws.[189] While database publishers need not contribute any intellectual achievement for which a reward is justifiable in terms of social costs, they have now staked a claim to subject matter that world intellectual property law had left unprotected as a building block of scientific and technological progress.

In seeking an unprecedented level of state intervention, therefore, it seems only logical that publishers should exchange a measure of support for the public-good uses of scientific data for lessened risk aversion and for a measure of artificial lead time in which to recoup their investments and turn a profit. This logic is reinforced by the fact that much, if not most, of the data likely to be commercialized under the proposed *sui generis* regime will, at some stage, have been a product of public-good undertakings funded largely by governments. Requiring publishers to further the public-good aspects of scientific data hardly seems unreasonable in this context, especially in view of the potential for rent-seeking that inheres in a market structure dominated by sole-source providers.

At the same time, one cannot push the concept of fair use to the point of requiring the private sector to make up for diminished government support of scientific research in general and of the generation of data in particular. Policymakers must, indeed, take pains to avoid a worst-of-both-worlds outcome, in which government support for the production of scientific data declines, while private investment in the generation, distribution, and application of data languishes for lack of adequate incentives. To the extent that private industry develops electronic information tools specifically to promote scientific investigation or other educational endeavors, the imposition of a subsidy favoring science becomes harder to justify and even counterproductive, given that scientific and educational institutions must pay for the many tools they use. This said, data nonetheless constitute a unique kind of tool, and no amount of investment can justify their greatly diminished availability for scientific inquiry.

Appropriate fair use provisions should thus be seen as part of a new cultural bargain that responds to serious concerns about the ability of data publishers to control access to scientific data as such. Implementing this bargain will require careful distinctions between uses that are "free" and those that providers must permit, but on fair and reasonable terms and conditions.

For example, scientists must freely be able to use the data underlying existing scientific theories to verify or challenge those theories and to develop new ones. Similarly, researchers should have completely free use of their own notes and working files in the conduct of their investigations, regardless of whether these files are embodied in electronic or print media. By the same token, a scientist who creates a new database while using another lawfully obtained database covered by a *sui generis* law, along with other data, should owe nothing but reasonable use or access charges to the proprietary rightholder if he or she did not reproduce a substantial component of the protected data in the new database. Indeed, a *sui generis* law should never prevent anyone, including scientists, from reproducing or using an insubstantial part of the contents of a protected database for virtually any lawful purpose.

Ascertaining fair uses that database owners must permit the scientific and educational communities to make on more favorable terms than those applicable to ordinary commercial users constitutes a more delicate task. As discussed throughout this report, whenever a given database is funded by government, a bedrock concept of fair use should require that the scientific and educational communities have access to its contents at no more than the marginal cost of reproduction and dissemination.[190]

In other words, data generated by public funds should come freighted with a built-in, cost-based discount for scientific research and education as a condition of further commercialization by others. This principle mainly preserves the status quo, at least for U.S. scientists, without shifting the costs of generating and distributing raw or processed scientific data onto private publishers. As long as enlightened government policy continues to favor substantial funding of the data-generating processes, this principle would promote science by preserving the public-good aspects of the data thus collected, without unduly inhibiting private incentives to invest. By the same token, it would prevent the private sector from displacing (or appropriating) the public-good aspects of government-funded data merely because *sui generis* legislation had been enacted to stimulate investment in distribution or value-added applications. While such a policy may conflict with the 1996 European Directive, depending on how the European Union member states choose to implement the relevant provisions, its adaptation in the United States could influence other countries, including even some European Union member states, which might decide to exercise their implementing option in precisely this way.[191]

Conversely, when the private sector or other nongovernmental entities fund the generation or distribution of data, a different fair use calculus should come into play. Here the problem is that the ability of science to pay the going, commercial rates is not commensurate with its resources or with the public interest in a strong, basic scientific establishment. The solution is not to shunt the problems of science onto publishers, who have their own business risks to manage, but to ensure that publishers charge scientific and educational users fair and reasonable prices that take account of the overriding public interest at stake.

Achieving this goal, however, is complicated by the difficulties of weaning sole-source providers from the rent-seekers' mentality if market forces themselves do not compel more favorable treatment of scientific and educational users. The appropriate response is to incorporate legal standards into the database law that can create sufficient leverage for scientific and educational users to obtain such treatment. The gentlest and least market-distorting form of leverage, in turn, is the legal uncertainty with which legislatures can endow the relevant fair use provisions. This strategy gives both sides the maximum incentives to negotiate their own licenses providing for price discrimination, product differentiation,[192] and other forms of relief on terms that seek to reconcile the different interests at stake.

Various technical devices, adopted singly or in combination, can be employed to bring about this result. For example, a general clause governing licenses in the database law can expressly provide that all licensing and distribution agreements effected under such a regime must be made "on fair and reasonable terms, with due regard for the needs of the scientific and educational communities, for the public interest in preserving competition, and for the needs of national economic development."[193] Such clauses, which have already been used in some database transactions,[194] would then be construed in the light of other provisions favoring publishers, so as not unduly to impair the commercial value of the database or the owner's return on his or her investment.[195] This approach should at least induce publishers to develop favorable subscription rates for academic and research institutions rather than insisting on per-use charges that may or may not apply in other circumstances.

The use of compulsory licensing can also increase the bargaining power of privileged users. As previously indicated, a properly crafted liability regime protecting investment in databases (which some investigators recommend) could itself incorporate an automatic license favoring second comers and value-adding users, which would kick in after an initial period of guaranteed lead time.[196] A refinement of this mechanism could then allow the scientific and educational communities to trigger a special compulsory license for essential needs in the event that publishers fail to provide reasonable terms and conditions.[197] If Congress ultimately adopts an exclusive rights regime for database owners, rather than an unfair competition model or a more refined liability model, such a regime could nonetheless include non-voluntary license provisions to meet the needs of these communities. In theory, such a clause permits either side to seek a judicial decision triggering, or blocking, the compulsory license for privileged uses. In practice, a built-in duty to negotiate before seeking such a license,[198] coupled with the uncertainty inherent in the applicable legal standards (and the well-known limits of judicial capability), should almost invariably lead to an accommodation between publishers and the research and educational communities that would remove the bone of contention.

Assessing Long-term Effects

Even if the research and educational communities persuade Congress to take their needs into account when fashioning a *sui generis* database law, their attention must remain focused on its operational consequences, assuming that the above-mentioned issues have been resolved. The introduction of new legal instruments, and a shift toward the commercialization of data, may profoundly change current institutions, especially those bearing on the funding of scientific research. The effects of these changes certainly will need to be assessed and monitored over time.

Where necessary, steps must be taken to ensure that new institutions suited to the maintenance of scientific progress are set in place before existing institutions are undermined or eliminated. The public funding of basic scientific data collection and dissemination should remain sufficiently robust to support the level of technological applications that has enabled U.S. firms to retain a competitive edge in the global marketplace. In this connection, if conciliatory efforts fail to dissuade foreign government science agencies from overcharging researchers and educators (i.e., charging more than marginal cost) for essential scientific data, countervailing pricing strategies and other tactics may become unavoidable. In any event, government agencies, the research and education communities, and database makers will have to cooperate at the implementation stage, with a view to reconciling the greater role of the private sector with the public-good aspects of national science policy.

From a long-term perspective, the research and education communities must face up to the fact that new technologies for generating, evaluating, and distributing data will continue to change many institutions on which basic science has traditionally relied. While certain to be disruptive, such changes, suitably guided, could foster and enhance vast new opportunities for the sciences, provided that the scientific community moves to meet the challenges in a timely and sustained fashion.

For example, new modes of organizing and distributing the funds needed to generate data may be devised, while the prospects for internalizing transmission and publication costs through the use of electronic communications networks merit careful study. To the extent that widespread commercialization or privatization of data does result from the adoption of new intellectual property rights and other factors, it could stimulate the scientific community to organize its own institutions or mechanisms for the management and dissemination of scientific data, which could operate outside the commercial arena. Because the research and educational communities are both producers and consumers of data, collective action along these lines could make science itself an increasingly important player in the market for databases generally, as well as a stabilizing force in determining the balance between public and private interests.[199] How to organize such large-scale undertakings will require careful thought and study in order to avoid sacrificing or compromising other goals of scientific endeavor, or undermining traditional norms of science emphasizing objective pursuit of knowledge based on full and open exchange of data.

Meanwhile, the adoption of different legal regimes to protect database makers by countries with different agendas and at varying stages of economic development could further complicate the full and open flow of scientific data across international frontiers. Measures to harmonize the domestic database-protection laws, or at least their effects on the transnational flow of scientific data, will therefore require intergovernmental study, as will measures and proposals affecting the regulation of national and global information infrastructures. Pressures to integrate these and other international intellectual property standards ever more deeply into the global trade apparatus will certainly mount as countries move to implement and expand the TRIPS Agreement and related international conventions within the framework of the World Trade Organization and that of WIPO, which continues to administer the Paris and Berne conventions.

The ensuing tensions and conflicts will make it more necessary than ever to develop a framework treaty to safeguard the full and open exchange of scientific data in an increasingly commercialized environment. The difficulty of regulating both public and private interests within such a treaty should not be underestimated, however, while the developing and least-developed countries are likely to play a more conspicuous role in intellectual property policymaking as time goes on.

RECOMMENDATIONS REGARDING LEGAL DEVELOPMENTS AFFECTING ACCESS TO DATA

The new proposals supporting an overly protectionist property rights regime for the contents of databases and for on-line transmissions of data and other scientific information have reached an advanced stage of legislative consideration at both the national and the international levels. The committee believes that these legislative changes do not reflect adequate consideration of the potential negative impacts on scientific research and education and that they have been proposed for implementation at an unnecessarily precipitous pace. The committee therefore recommends that the Office of Science and Technology Policy, leaders from the science agencies and professional societies, and all those concerned with sustaining the health of the scientific enterprise should immediately take the following actions:

1. Present to all relevant legislative forums the principle of full and open exchange of scientific data resulting from publicly funded research, and clarify the importance of sustaining such exchange to the nation's future whenever these forums consider laws that would apply to exchange of scientific data.

2. Demand that national and international legislative processes now in progress slow to a rational pace, and that the deliberations become more public to allow the scientific and educational communities to present their views and concerns to lawmakers.

3. Advocate the incorporation of equivalents of "fair use" as part of any regulatory structure applying to databases as such, or to on-line storage and transmission of data and other scientific information. As a corollary, ensure that the public-good aspects of scientific data are preserved and promoted in laws and regulations governing intellectual property on the Internet and in any future electronic networked environments.

4. Work with Congress and the official U.S. representatives to the World Trade Organization and the World Intellectual Property Organization to ensure that the nation's interests in maintaining preeminence in science and technology are not undermined.

5. Pursue these issues not only within the United States, but also internationally through international scientific organizations and U.S. foreign-policy channels as they deal with trade and other agreements affecting intellectual property protection.

NOTES

1. For clarification and further documentation of legal issues discussed in this chapter, see the article by J.H. Reichman and Pamela Samuelson (1997), "Intellectual Property Rights in Data?," *Vanderbilt Law Review*, 50(1):51-166.
2. U.S. Constitution, Article I, Section 8, clause 8.
3. See, e.g., Office of Technology Assessment (1986), *Intellectual Property Rights in an Age of Electronics and Information*, U.S. Government Printing Office, Washington, D.C., pp. 59-97; Anne Wells Branscomb (1994), *Who Owns Information? From Privacy to Public Access*, Basic Books, pp. 1-8.
4. See, e.g., Commission of the European Communities (1988), "Green Paper on Copyright and the Challenge of Technology—Copyright Issues Requiring Immediate Action," June 7 (hereinafter E.C. Green Paper); Laurence Kay (1995), "The Proposed E.U. Directive for the Legal Protection of Databases: A Cornerstone of the Information Society?," *Eur. Intell. Prop. Rev.* (hereinafter *EIPR*) 17:583.
5. World Intellectual Property Organization (WIPO) (1996) "Basic Proposal for the Substantive Provisions of the Treaty on Intellectual Property in Respect of Databases to be Considered by the Diplomatic Conference" (Memorandum prepared by the Chairman of the Committee of Experts), WIPO doc. CRNR/DC/6, Geneva, August 30 (hereinafter WIPO Draft Database Treaty).
6. Both international copyright law under the Berne Convention and the domestic laws of most developed countries require that authors of literary works obtain a "moral right" to proper attribution for their published creations. See, e.g., Berne Convention, article 6 *bis*. Nevertheless, the U.S. Congress has so far declined to implement this obligation except with regard to visual or graphic artists (see 17 U.S.C. 106A (1994)), which leaves the problem of attribution largely to the vagaries of state and federal unfair competition law.
7. See, e.g., Frank Gotzen (1987), "Grands Orientations du Droit d'Auteur dans les Etats Membres de la C.E.E. en Matière de Banques de Données," *Banques de Données et Droit d'Auteur*, pp. 85-98 (IRPI ed.); Paul Edward Geller (1991), "Copyright in Factual Compilations: U.S. Supreme Court Decides *Feist* Case," *Int. Rev. Ind. Prop. Copyright L.* (IIC) 22:802; Erbert J. Dommering (1991), "An Introduction to Information Law: Works of Fact at the Crossroads of Freedom and Protection," *Protecting Works of Fact: Copyright, Freedom of Expression, and Information Law*, 1-40 (E.J. Dommering and P. Bernt Hugenholz, eds.). However, the Nordic countries constituted an interesting exception by instituting *sui generis* laws to protect noncopyrightable compilations at a relatively early period. See, e.g., Gunnar W.G. Karnell (1991), "The Nordic Catalogue Rule," *Protecting Works of Fact*, pp. 67-72.
8. See, e.g., J.H. Reichman (1995), "Charting the Collapse of the Patent-Copyright Dichotomy: Premises for a Restructured International Intellectual Property System," *Cardozo Arts & Entertain. L. J.*, 13:475.
9. See, e.g., Jessica Litman (1990), "The Public Domain," *Emory L. J.*, 39:965, 1015 ("Giving an author a copyright in . . . a basic building block of her art . . . risks denying that basic building block to all other authors who come into even fleeting contact with the first author's work").
10. See Reichman and Samuelson, note 1, at pp. 58-72, 137 (citing authorities).
11. See, e.g., *Kewanee Oil Co. v. Bicron Corp.*, 416 U.S. 470 (1974); Uniform Trade Secrets Act, 14 ULA §1(4) (1985). See generally J.H. Reichman (1992), "Overlapping Proprietary Rights in University-Generated Research Products: The Case of Computer Programs," *Columbia-VLA J. Law & Arts*, 17:51, 93-98.
12. Legal liability attached only to third parties who engaged in improper means of reverse engineering, such as bribing employees or resorting to industrial espionage. In such cases, the free-riding offender was required to compensate the innovator only for the lost profits that would have accrued during the hypothetical period that would have been required to reverse engineer the product in question by honest means.

13. See, e.g., Paul Goldstein (1996), *Copyright: Principles, Law and Practice*, §1.2, 2d ed., Little, Brown; William Kingston (1990), *Innovation, Creativity and Law*, Klüwer Academic Publishers, Dordrecht, pp. 83-85; see also Ejan Mackaay (1982), *Economics of Information and Law*, Klüwer-Nijhdt, pp. 115-17 (noting market distortions ensuing from public good problems and uncertainties that would require a higher expected return).
14. See, e.g., 17 U.S.C. §202 (1994) (distinguishing ownership of copyrightable work from ownership of material support); J.H. Reichman (1989), "Intellectual Property in International Trade: Opportunities and Risks of a GATT Connection," *Vand. J. Transnat'l L.*, 22:747, 800-06.
15. See, e.g., Peter A. Jaszi (1996), "Goodbye to All That—A Reluctant (and Perhaps Premature) Adieu to a Constitutionally-Grounded Discourse of Public Interest in Copyright Law," *Vand. J. Transnat'l L.*, 29:595, 599-600 (stressing "economic and cultural bargain between authors and users . . . at the heart of U.S. [copyright] law, as reflected in the Patent and Copyright Clause [of the Constitution], and a parade of Supreme Court precedents"). See also Robert A. Kreiss (1995), "Accessibility and Commercialization in Copyright Theory," *UCLA L. Rev.* 43:1, 6-22; J.H. Reichman (1993), "Electronic Information Tools—The Outer Edge of World Intellectual Property Law," *Int. Rev. Indus. Prop. Copyright L.*, 25:446, 461 (stressing that "these fictitious portable fences neutralize essential attributes of property that possession would ordinarily confer").
16. See 17 U.S.C. §102(a), (b); *Feist Publications, Inc. v. Rural Telephone Service Co.*, 111 S. Ct. 1282 (1991).
17. See, e.g., 17 U.S.C. §108 (reproduction by libraries and archives), 109(a) (first-sale doctrine), 110(a) (face-to-face teaching activities), 110(b) (broadcasts of nondramatic literary or musical works for certain educational purposes), 114 (limiting rights and scope of protection in sound recordings), 115 (compulsory license for musical works recorded on sound recordings), 117 (archival uses of computer programs), 118 (exemptions for use by noncommercial broadcasters), 120 (right to photograph architectural works) (1994).
18. 17 U.S.C. §107 (1994).
19. See, e.g., 17 U.S.C. §107 (1994); *Campbell v. Acuff-Rose Music, Inc.*, 114 S. Ct. 1164 (1994); *Harper & Row Publishers, Inc. v. Nation Enterprises*, 471 U.S. 539 (1985).
20. See, e.g., *American Geophysical Union v. Texaco*, 60 F.3d 913 (2d. Cir. 1995); *Sega Enterprises Ltd. v. Accolade, Inc.*, 977 F. 2d 1510 (9th Cir. 1992); Pamela Samuelson (1993), "Fair Use for Computer Programs and Other Copyrightable Works in Digital Form: The Implications of *Sony*, *Galoob*, and *Sega*," *J. Intell. Prop. L.*, 1:49.
21. See, e.g., Paul Goldstein (1994), *Copyright's Highway—From Guttenberg to the Celestial Jukebox*, pp. 129-30 (stating that "the risk has grown that 'private' copies will displace the retail sales and rentals of the authorized originals from which publishers, record companies, and motion picture producers earn their revenues"). The extent to which private photocopying of journals for research purposes without compensation remains a fair use is controversial from a global perspective. See, e.g., William R. Cornish (1994), "Copyright in Scientific Works (Scientific Communications, Computer Software, Data Banks): An Introduction," *European Research Structures—Changes and Challenges: The Role and Function of Intellectual Property Rights*, pp. 47-50 (Max-Planck-Gesellschaft, ed.) (stressing that academics "who are also authors, find their interests . . . and . . . judgment . . . pulled in two directions on these issues" and no settled solution has emerged); Jane C. Ginsburg (1992), "Reproductions of Protected Works for University Research or Teaching," *J. Copyright Soc'y* 39:181, 188-89, 192-211 (discussing legal license regimes and collection societies). But see L. Ray Patterson and Stanley W. Lindberg (1991), *The Nature of Copyright*, pp. 191-196 (arguing that private use for nonprofit purposes is always allowed).
22. See, e.g., Wendy J. Gordon (1982), "Fair Use as Market Failure: A Structural and Economic Analysis of the Betamax Case and Its Predecessors," *Colum. L. Rev.*, 82:1600. For recent tensions with this theory, compare, e.g., P. Goldstein, note 13, §10.1 (evaluating public and

private benefits test of fair use in terms of transaction costs) with Goldstein, *Celestial Jukebox*, note 21, at pp. 223-24 ("celestial jukebox may reduce transaction costs" and "perceived need for safety valve such as fair use," which would lead to a copyright law with "no exemptions from liability"). But see Goldstein, *Celestial Jukebox* at p. 230 (stressing the enduring importance of "exemptions or compulsory licenses for educational and research uses").

23. See, e.g., *Feist*, 111 S. Ct. at 1292 (stressing adverse effects on free flow of information by "creat[ing] monopol[ies] in public domain materials"); *Harper & Row v. Nation Enterprises*, 471 U.S. 539, 555 (1985), stressing First Amendment values in free flow of factual information; *Financial Information, Inc. v. Moody's Investor's Serv., Inc.*, 808 F. 2d 204, 207 (2d Cir. 1986), cert. denied, 484 U.S. 820 (1987) (stressing "risk [of] putting large areas of factual research material off limits and threatening the public's unrestrained access to information"). See also Philip H. Miller (1991), "Life After *Feist*: The First Amendment and the Copyright Status of Automated Databases," *Fordham L. Rev.* 60:507; Michael J. Haungs (1990), "Copyright of Factual Compilations: Public Policy and the First Amendment," *Colum. J. Law & Social Problems*, 23:347, 364; Robert C. Denicola (1981), "Copyright in Collections of Facts: A Theory for the Protection of Nonfiction Literary Works," *Colum. L. Rev.*, 81:516, 525. For the view that legal protection of facts and data as such is consistent with the First Amendment on certain conditions, such as the availability of noncommercial fair use and compulsory licenses, see, e.g., Jane C. Ginsburg (1992), "No 'Sweat'? Copyright and Other Protection of Works of Information After *Feist v. Rural Telephone*," *Colum. L. Rev.*, 92:338, 384-87.

24. See 17 U.S.C. §106(1), (2) (1994); *Feist Publications, Inc., v. Rural Telephone Service Co.*, 111 S. Ct. 1282, 1289-91 (1991); *Key Publications, Inc. v. Chinatown Today Publishing Enterprises, Inc.*, 945 F 2d 509 (1991); *Kregos v. Associated Press*, 937 F. 2d 700 (2d Cir. 1991); *Victor Calli Enterprises, Inc. v. Big Red Apple, Inc.*, 936 F. 2d 671 (2d Cir. 1991); *BellSouth Advertising & Publishing Corp. v. Donnelly Information Publishing, Inc.*, 999 F. 2d 1436 (11th Cir., en banc, 1993), cert. denied, 114 S. Ct. 943 (1994). If courts strictly apply *Feist* at both the eligibility and the scope of protection phases, and thus continue to reject stronger protection based on "sweat-of-the-brow" investment theories that some jurisdictions had embraced prior to *Feist*, the effect is to "strip . . . away or sharply reduce . . . the copyright protection afforded a variety of 'information products,' from directories and mailing lists to computerized databases." Ginsburg, "Information After Feist," note 23, at p. 339. See also Denicola, note 23 (advocating compiler's copyright to overcome lack of incentives); Jane C. Ginsburg (1990), "Creation and Commercial Value: Copyright Protection of Works of Information," *Colum. L. Rev.*, 90:1865 (advocating copyright protection of low-authorship factual works, including databases, but proposing compulsory license for derivative users of data. For comparative law, see especially Alain Strowel (1993), *Droit d'Auteur et Copyright—Divergences et Convergences* (Bruglaut, Brussels) and L.G.D.J. (Paris), pp. 29-30, 264-66, 391-474.

25. See, e.g., *CCC Info. Servs. Inc. v. Maclean Hunter Market Reports, Inc.*, 44 F. 3d 61 (2d Cir. 1994); *Lipton v. Nature Co.*, 71 F 3d 464 (2d Cir. 1995); *Warren Publishing, Inc. v. Microdos Data Corp.*, 52 F. 3d 950 (11th Circ. 1995).

26. See, e.g., Ginsburg, "Information After *Feist,"* note 23, at pp. 367-73; Wendy J. Gordon (1992), "On Owning Information: Intellectual Property and the Restitutionary Impulse," *Va. L. Rev.*, 78:149; Dennis S. Karjala (1994), "Misappropriation as a Third Intellectual Property Paradigm," *Colum. L. Rev.*, 94:2594. But see Leo J. Raskind (1991), "The Misappropriation Doctrine as a Competitive Norm of Intellectual Property Law," *U. Minn. L. Rev.*, 75:875.

27. See, e.g., Reichman, "Collapse of the Patent-Copyright Dichotomy," note 8, at pp. 512-17, 513, n. 176; also, quoting J.H. Reichman (1994), "Legal Hybrids Between the Patent and Copyright Paradigms," *Colum. L. Rev.*, 94:2504, n. 401:

> Factors pulling for over- or underprotection already exist on both sides of the classical line of demarcation [between the patent and copyright subsystems]. On the copyright side . . . for example, a broad derivative work right sometimes overprotects by favoring overlapping claims to incremental

innovation while restricting access to ideas, methods and processes by indirect means and for a very long duration . . . Yet, underprotection can result from the inability of copyright-like models to protect the internal dynamic features of technological innovation, in which idea and expression merge, and also from the lack of any exclusive right to control end use . . . Similarly, on the industrial property side . . ., "overprotection results from the progressive monopolization of ever smaller aggregates of inventive activity, which elevate social costs in return for no clearly equilibrated social benefits. Yet, the nonobviousness standard [of patent law] and its variants can also induce states of chronic underprotection by excluding the bulk of the incremental innovations that underlie today's most promising technologies.

28. See, e.g., Debra B. Rosler (1995), "The European Union's Proposed Directive for the Legal Protection of Databases: A New Threat to the Free Flow of Information," *High Tech. L. J.*, 10:105; Paul Durdik (1994), "Ancient Debate, New Technology: The European Community Moves to Protect Computer Databases," *Boston Univ. Int'l L. J.*, 12:153, n. 1 (citing E.C. sources showing that U.S. and European producers account, respectively, for about 56 percent and 25 percent of "world electronic information services output," but that the European share had been "only one-tenth the size of the U.S. market" in 1980. For the view that U.S. dominance of the private sector stems from its loose copyright regime, which "most closely approximates a free market for data compilations" based on superior product development, see Charles von Simson (1995), "*Feist* or Famine: American Database Copyright as an Economic Model for the European Union," *Brooklyn J. Int'l L.*, 20:729.
29. See, e.g., Rosler, note 28. For evidence of a similar phenomenon with regard to databases that serve the legal community, see, e.g., *West Publishing Co. v. Mead Data Central, Inc.*, 799 F. 2d 1219 (8th Cir. 1986), *cert. denied*, 479 U.S. 1070 (1987); see also L. Ray Patterson and Craig Joyce (1989), "Monopolizing the Law: The Scope of Copyright Protection for Law Reports and Statutory Compilations," *UCLA L. Rev.*, 36:719.
30. See, e.g., 17 U.S.C. §105 (1994) ("Copyright protection under this title is not available for any work of the United States government." A "work of the United States government" is a "work prepared by an officer or employee of the U.S. government as part of that person's official duties." 17 U.S.C. §101 (1994)).
31. Nevertheless, some disputes are likely to be adjudicated, owing to the uncertainties that arise from the application of overlapping legal regimes to new technologies and to the gaps or ambiguities in internal university legislation. See, e.g., Rochelle Cooper Dreyfuss (1987), "The Creative Employee and the Copyright Act of 1976," *U. Chi. L. Rev.*, 54:590, 597-98, 640-41; Reichman, "Overlapping Proprietary Rights," note 11.
32. Leslie A. Kurtz (1996), "Copyright and the National Information Infrastructure in the United States," *EIPR*, 18:120. See also Pamela Samuelson, "Technological Protection for Copyrighted Works," paper presented to the Thrower Symposium, Emory Law School, Feb. 22, 1996 (stating that, although digital technology "poses a serious challenge for copyright owners because works in digital form are vulnerable to uncontrolled replication and dissemination in networked environments," it is "not just part of the problem; it may also be part of the solution").
33. Gregory M. Hunsuker (1997), "The European Database Directive: Regional Stepping Stone to an International Model?," *Fordham Intell. Prop., Media & Entertain. L.*, forthcoming.
34. Hunsuker, note 33. See also Pamela Samuelson, "Missing Foundations of the Proposed European Database Directive," paper presented to the Specialist Meeting on Law, Information Policy, and Spatial Databases, sponsored by the National Center for Geographic Information and Analysis, Arizona State University College of Law, Oct. 28-30, 1994 (stressing importance of add-on or value-added products in the digital environment).
35. See, e.g., Paul A. David and Dominique Foray (1995), "Accessing and Expanding the Science and Technology Knowledge Base," *Science, Technology, Industry Rev.* 16(3):38-59; and Paul Ginsparg (1995), "Winners and Losers in the Global Research Village," paper presented to UNESCO Conference on Electronic Publishing, February 19-23, 1996, Paris; see also, Cristiano

Antonelli (1992), "The Economic Theory of Information Networks," in *The Economics of Information Networks* (C. Antonelli, ed.), Elsevier Science Publishing Co.
36. See, e.g., David and Foray, note 35; see also Cristiano Antonelli, note 35, at pp. 5-28.
37. See, e.g., Hunsuker, note 33, at p. 1 (quoting sources that estimate the value of the global information industry to reach $3 trillion by early in the next century); W. Joseph Melnick (1994), "A Comparative Analysis of Proposals for the Legal Protection of Computerized Databases: NAFTA vs. the European Communities," *Case Western Reserve J. Int'l. L.*, 26:57, 59, n.14 (quoting sources that estimate E.C. database market at $10.2 billion, which amounted to about 30 percent of the world market in 1994).
38. See, e.g., E.C. Green Paper, note 4; National Information Infrastructure Task Force (1995), *Report of the Working Group on Intellectual Property and the National Information Infrastructure* (Sept.) (hereinafter U.S. White Paper). In a larger perspective, however, it has been argued that the legal problems of electronic databases are assimilable to those of industrial designs, computer programs, plant varieties, biogenetically engineered products, and numerous other forms of design-dependent, subpatentable innovation that fall into a widening penumbra between the increasingly obsolete patent and copyright paradigms. See, e.g., Reichman, "Legal Hybrids," note 27 (proposing a third intellectual property paradigm for investors in subpatentable innovation based on liability principles); Pamela Samuelson, Randall Davis, Mitchell Kapor, and J.H. Reichman (1994), "A Manifesto Concerning the Legal Protection of Computer Programs," *Colum. L. Rev.*, 94:2308 (proposing such a regime for computer programs).
39. See, e.g., *CCC Intro. Servs. Inc. v. Maclean Hunter Market Reports, Inc.*, 44 F. 3d 61 (2d Cir. 1994) (CCC took "virtually the entire compendium" of Maclean's used car valuations and "effectively offers to sell its customers Maclean's Red Book through CCC's database"); *Warren Publishing, Inc. v. Microdos Data Corp.*, 52 F 3d 950 (11th Cir. 1995) (statistically, Microdos' work contained from 96 to 99 percent of Warren's data on nationwide cable TV services).
40. See, e.g., Office of Technology Assessment (1992), *Finding a Balance: Computer Software, Intellectual Property, and the Challenge of Technological Change*, U.S. Government Printing Office, Washington, D.C., pp. 166-79 (deploring contracts in which software licensors obliged libraries to abrogate "rights described in the copyright law," among other practices); OTA Report, note 3, at p. 163; Cornish, note 21, at p. 50 (despite case for a measure of free reprography for purposes of academic research, "academic institutions are regarded as relatively soft targets by publishing interests [in U.K.], which have looked at them as suitable points for inserting initial wedges"). See generally Reichman, "Electronic Information Tools," note 15, at pp. 461-68 ("Public Interest at Odds with the Two-Party Deal"); Marshall Leaffer (1994), "Engineering Competitive Policy and Copyright Misuse," *U. Dayton L. Rev.*, 19:1087, 1094.
41. See the introductory discussion in Chapter 3.
42. See, e.g., Samuelson et al., "Manifesto," note 38; see also Joseph Straus and Rainier Moufang (1990), *Deposit and Release of Biotechnological Material for the Purpose of Patent Procedure*, Nomos Verlags Gesellschaft (discussing biotechnical research and development), and Dan L. Burk (1991), "Biotechnology and Patent Law: Fitting Innovation to the Procrustean Bed," *Rutgers Comp. Tech. L. J.*, 17:1, 33-34.
43. See, e.g., John Browning (1996), "Cyber View: Playing Facts and Loose," *Scientific American* (June): 30-32 (warning about unintended effects of legal restrictions on searching and gathering data).
44. Charles von Simson (1995), "*Feist* or Famine: American Database Copyright as an Economic Model for the European Union," *Brooklyn J. Int. L.*, 20:729, 768.
45. If the second comer independently generates its own data, or combines its inputs with the first comer's data to produce value-adding applications, the former contributes knowledge, capital, and skilled efforts to the data-generating communities' overall endeavor. These second com-

ers, who do not merely duplicate or "clone" the first comer's product, are hardly free riders even when they do not contribute directly to the first comer's production costs under a licensed royalty transaction. See, e.g., Samuelson, "Missing Foundations," note 34; see generally Reichman, "Legal Hybrids," note 27, at pp. 2521-23, 2535-39.
46. See Reichman and Samuelson, note 1, at pp. 66-70, 124-30.
47. In one recent case, for example, a database maker spent about $10 million to compile some 95,000,000 residential and commercial listings from some 3,000 telephone directories. A purchaser who paid $200 for a compact disk electronically extracted and recompiled part of the data and then made his listings available over the Internet. In *Pro CD Inc. v. Zeidenberg*, 908 F. Supp. 640 (W.D. Wis. 1996), the federal district court rejected the plaintiff's copyright claim as well as state law claims in contract and unfair competition law (see also Hunsuker, note 33, at pp. 13-14), but was reversed on appeal, in *ProCD, Inc. v. Zeidenberg*, 86 F. 3d 1447 (7th Cir. 1996).
48. Jessica Litman (1992), "After Feist," *U. Dayton L. Rev.*, 17:607, 611 (adding that, "[i]ndeed, a large number of on-line database . . . [publishers] availed themselves of those strategies well before the *Feist* decision").
49. See Reichman, "Electronic Information Tools," note 15, at pp. 461-67 ("Public Interest at Odds with the Two-Party Deal"); Jane C. Ginsburg (1993), "Copyright Without Walls?: Speculations on Literary Property in the Library of the Future," *Representation*, 42 (Spring):53, 60-63.
50. See, e.g., Goldstein, *Celestial Jukebox*, note 21, at pp. 223-24; Jessica Litman (1994), "The Exclusive Right to Read," *Cardozo Arts & Entertain. L. J.*, 31:29, 32; Samuelson, "Technological Protection," note 32.
51. See note 3 reporting OTA concerns in this regard; Kurtz, note 32.
52. See Directive 96/9/E.C. of the European Parliament and of the Council of March 11, 1996, on the legal protection of databases, 39 O.J.L.77/20, March 27, 1996 (hereinafter E.C. Directive on Databases or Final E.C. Directive); see also E.C. Council of Ministers, Common Position No. 20/95, adopted 10 July 1995, with a view to adopting Directive 95//E.C. of the European Parliament and of the Council . . . on the legal protection of databases, 95/C288/02, O.J.C.288/14 (Oct. 10, 1995) (hereinafter E.C. Common Position).
53. See H.R. 3531, U.S. Congress, House of Representatives, 104th Cong., 2d Sess., May 23, 1996, Database Investment and Intellectual Property Antipiracy Act of 1996.
54. See WIPO Draft Database Treaty, note 5, article 5 (I) (proposing a truncated version of the Berne Convention's fair use clauses).
55. Cf., e.g., Kreiss, note 15, at pp. 33-34; Mark A. Lemley (1995), "Intellectual Property and Shrink-wrap Licenses," *So. Cal. L. Rev.*, 68:1239-1294; Leaffer, note 40; David A. Rice (1992), "Public Goods, Private Contract and Public Policy: Federal Preemption of Software License Prohibitions Against Reverse Engineering," *U. Pitt. L. Rev.*, 53:543. But see 17 U.S.C. §108(1)(4) (1994) (allowing contractual obligations to override specified library privileges); Ginsburg, "Speculations," note 49.
56. See, e.g., Jaszi, note 15, at p. 599 (criticizing replacement of "cultural bargain" theory of copyright law with new, trade-driven goal, which seeks to "enhance . . . the wealth and overall financial well-being of companies which invest in the production of and distribution of copyrighted works"); see also David Nimmer (1995), "The End of Copyright," *Vand. L. Rev.*, 48:1385.
57. Ginsburg, "Speculations," note 49, at p. 60.
58. See, e.g., Ginsburg, "Speculations," note 49, at pp. 60-63 (arguing for freedom of publishers to condition libraries' access to digitally delivered information on compliance with a variety of restrictions, regardless of principles such as fair use, because information providers need not resort to libraries as conduits for digital information in the future except, perhaps, as "full-

service help-line"). But see Jane C. Ginsburg (1994), "Surveying the Borders of Copyright," *J. Copyright Soc'y,* 41:322: 325-26 (arguing for legal restraints on such contractual conditions).

59. See, e.g., Goldstein, *Celestial Jukebox,* note 21, at p. 230 ("Exemptions and compulsory licenses for research and educational uses recognize the transcendent claim these uses have on a copyright system whose founding premise is that a culture can be built only if toilers in the vineyard are free to draw on the works of their predecessors"); Marci A. Hamilton (1996), "The TRIPS Agreement: Imperialistic, Outdated and Overprotective," *Vand. J. Transnat'l L.,* 27:613, 623-33 (advocating construction of "free-use zone . . . in the online era"); Samuelson, *Technological Protection,* note 37. For the view that developing countries should formulate their own doctrines of misuse to govern information providers' contracts, see J.H. Reichman (1997), "From Free Riders to Fair Followers: Global Competition Under the TRIPS Agreement," *NYU J. Int'l L. Pol.* (forthcoming).

60. See, e.g., Hamilton, note 59, at pp. 628-29.

61. See, e.g., Goldstein, *Celestial Jukebox,* note 21, at p. 230 (stressing need for exemptions and compulsory licenses favoring "research and educational uses" as transcendent claim rooted in cumulative progress of knowledge).

62. As mentioned above, the copyright laws of most developed countries exclude functionally determined databases and do not protect disparate data even when a given compilation as a whole happens to satisfy the eligibility requirements of those laws. This principle was incorporated into Article 10 of the TRIPS Agreement of 1994, which requires copyright protection when "the selection or arrangement of . . . contents constitute intellectual creations," but stipulates that such protection "shall not extend to the data or material itself."

63. See Paris Convention for the Protection of Industrial Property, March 20, 1883, as last revised at Stockholm, July 14, 1967, 21 U.S.T. 1583, T.I.A.S. No. 6923, 828 U.N.T.S. 305 (hereinafter Paris Convention); Berne Convention for the Protection of Literary and Artistic Works, Sept. 9, 1886, as last revised at Paris, July 24, 1971, S. Treaty doc. 99-27, 99 Cong. 2d Sess. (1986), 828 U.N.T.S. 221 (hereinafter Berne Convention). For the official line of demarcation between "writings" and "products" that underlies these Conventions, and its gradual disintegration under pressure from a proliferating set of hybrid (i.e., *sui generis*) regimes that deviate from the patent and copyright models, see, e.g., Reichman, "Collapse of the Patent-Copyright Dichotomy," note 8, at pp. 480-512; see also Reichman, "Legal Hybrids," note 27, at pp. 2448-2519.

64. While the Commission claims that a key motive is the need to harmonize European Union law, critics debunk this claim because Article 10 of the TRIPS Agreement partly performed this function, and also because the E.C.'s database regime, as finally adopted, actually discourages harmonization on the crucial issue of fair use. See, e.g., Charles R. McManis (1996) "Taking TRIPS on the Information Superhighway: International Intellectual Property Protection and Emerging Computer Technology," *Villanova L. Rev.,* pp. 207-288. The predominant objective, among those stated, is to increase the share of European database producers (including governments) in the world market. See, e.g., E.C. Directive on Databases, note 52, Recital 11.

65. See, e.g., Commission of the European Communities (1991), *1991 Report on the Impact Program: Main Events and Developments in the Electronic Information Services Market,* COM (93) 156 final; Commission of the European Communities (1990), *Working Program of the Commission in the Field of Copyright and Neighboring Rights,* COM (90) 584 final; Rosler, note 28, at pp. 105, 107, 110-13. The IMPACT program specifically addressed the goal of improving the position of the European Union's member countries in the emerging global market for information goods. Among the strategies it endorsed were proposals to strengthen intellectual property rights, to protect new technologies, and to stimulate both international trade and European economic development.

66. See, e.g., Rosler, note 28, at pp. 109-10, 133-39. The Commission stressed the vulnerability of database publishers to market failure, but devoted little or no published attention to the

countervailing risk of technologically induced overprotection. It nonetheless attempted to deal with this latter problem by means of a compulsory license, but was foiled by the Council of Ministers at the last moment.
67. See E.C. Directive on Databases, note 52, articles 3-6 (copyright), 7-11 (*sui generis* right). For earlier versions, see Commission of the European Communities (1992), *Proposal for a Council Directive on the Legal Protection of Databases*, COM (92) 24 final—SYN 393 (First E.C. Directive on Databases); Commission of the European Communities (1993), *Amended Proposal for a Council Directive on the Legal Protection of Databases*, COM (93) 464 final—SYN 393 (Amended E.C. Directive on Databases); Council of the European Communities, Common Position, note 52.
68. See, e.g., Gunnar Karnell (1991), "The Nordic Catalogue Rule," *Protecting Works of Fact*, pp. 67-72 (E.J. Dommering and P.B. Hugenholtz, eds.), Klüwer. Laws implementing this regime "prohibit slavish reproduction, in whole or in part, of 'catalogues, tables, and similar compilations in which a large number of particulars have been summarized' including databases, for ten years after first publication ... [I]ndustrious effort and investment rather than creativity are the prerequisites." Reichman, "Legal Hybrids," note 27, at pp. 2492-93 (quoting Karnell and noting predigital ambiguities of this law).
69. Reichman, "Legal Hybrids," note 27, at p. 2493; see also Jean Hughes and Elizabeth Weightman (1992), "E.C. Database Protection: Fine Tuning the Commission's Proposal," *EIPR*, 14:146, 148 (Directive goes beyond the Nordic rule and protects against reuse of the data compiled).
70. See generally Common Position, note 52 at pp. 14-29.
71. See E.C. Directive on Databases, note 52.
72. See Haungs, note 23; Miller, note 23; and Rosler, note 28, and accompanying text. See also Ginsburg, "Information After Feist," note 23; Litman, "Public Domain"; Melville B. Nimmer and David Nimmer (1996), *Nimmer on Copyright*, §§1.10[C][2]. 1.10 [D], Matthew Bender.
73. See First E.C. Directive on Databases, articles 1(1), 2(5); Commission of the European Communities (1992), *Explanatory Memorandum to the Proposal for a Council Directive on the Legal Protection of Databases*, COM (92) 24 final—SYN 393, at pp. 21-22, 25, 35, 41) (hereinafter First Explanatory Memorandum); Amended E.C. Directive on Databases, articles 2.2, 6 (all stressing the goal of protecting the compiler's industrious effort and investment against parasitic appropriation by competitors). A true liability regime does not bestow winner-take-all rewards in exchange for certain technical achievements. Rather, a liability regime (such as the model trade secret law used in the United States) aims primarily to restore and preserve the bases for healthy competition by discouraging certain market-distorting forms of conduct that prevent innovators from appropriating the fruits of their investment. As long as innovators obtain adequate lead time and second comers contribute directly or indirectly to the innovators' costs of research and development, a liability regime declines to endow these same innovators with an absolute right to control the uses of their innovative products or with any legal barriers to entry by others. See, e.g., Reichman, "Legal Hybrids," note 27, at pp. 2434-42, 2496, 2504-2558.
74. See, e.g., Samuelson, "Missing Foundations," note 34; Rosler; note 28.
75. Now Preambular Recitals 6-7; First Explanatory Memorandum, note 73.
76. Cf. Uniform Trade Secrets Act §1(4), 14 U.L.A. 438 (1985) (UTSA) (adopted by a majority of states); Restatement (Third) of Unfair Competition Law §§39-45 (1993); David D. Friedman, William Landes, and Richard Posner (1991), "Some Economics of Trade Secret Law," *J. Econ. Persp.* (Winter), at 61 et seq.; Wendy J. Gordon (1992), "On Owning Information: Intellectual Property and the Restitutionary Impulse," *Virginia L. Rev.*, 78:149. See also *International News Service V. Associated Press*, 248 U.S. 215 (1918).
77. See First E.C. Directive on Databases, note 67, articles 1(1), 2(5), 9(3); Technically, the right arises with the creation of the database and lapses 10 (now 15) years from the date it was first lawfully made available to the public. The provision forbidding unauthorized reuse of the

compiler's factual contents closed a gap in the Nordic catalogue rules, which case law had not yet resolved. See notes 68 and 69 and accompanying text. Already at this first draft stage, however, the language chosen to implement the Commission's "unfair competition" approach was contradicted by other language describing the database maker's "exclusive right to prevent unauthorized extraction and reutilization" of contents. See, e.g., First Explanatory Memorandum, note 73, at p. 53.

78. See First E.C. Directive on Databases, note 67, articles 8(1), (2); see also Amended E.C. Directive on Databases, note 67, article 11(1), (2). Of course, if multiple data providers serviced a given market segment, the draft Directive's procompetitive thrust was satisfied without recourse to a compulsory license. However, the opportunity to choose among providers seems to be rare in practice because the bulk of all electronic compilations of data reportedly emanates from sole-source providers, and this "niche" marketing appears characteristic of both the private and the public sector. In all such cases, the compulsory license would lie, and originators, including public bodies benefiting from a natural monopoly, would be obliged to grant licenses for commercial reexploitation of their data on fair and nondiscriminatory terms.

79. In this respect, the early draft version seems to have anticipated some of the findings concerning the procompetitive characteristics of liability-based intellectual property regimes that legal theory was investigating at about the same period of time. See, e.g., Samuelson et al., "Manifesto," note 38, at p. 2308; Reichman, "Legal Hybrids," note 27, at p. 2432.

80. See Commission of the European Communities, *Amended Proposal for a Council Directive on the Legal Protection of Databases*, COM (93) 464 final—SYN 393, at p. 4 (1993) (hereinafter Second Explanatory Memorandum) (declining to accept E.U. Parliament's request for special exemptions in favor of education and research).

81. See, e.g., the discussion in Chapter 4.

82. See Amended E.C. Directive on Databases, note 67; Reichman, "Legal Hybrids," note 27, at pp. 2494-98 (analyzing and criticizing these proposals).

83. See E.C. Common Position, note 52; Hunsuker, note 33 (approving this version); von Simson, note 44 (criticizing this version); Reichman and Samuelson, note 1 (criticizing this version).

84. See E.C. Directive on Databases, note 52, articles 7(1) ("Member States shall provide for a right for the maker of a database which. . ."), 16(1) (requiring Member States "to comply with this Directive before 1 January 1998").

85. E.C. Directive on Databases, note 52, article 1(2) (broadly defining a database as "a collection of independent works, data or other materials arranged in a systematic or methodical way and individually accessible by electronic or other means").

86. See text accompanying notes 76 and 77.

87. E.C. Directive on Databases, note 52, article 7(1).

88. See, id., article 1(1) at 24 ("This Directive concerns the legal protection of databases in any form"). Both the First Proposed Directive, note 67, article 1(1), and the Amended E.C. Directive on Databases of 1993, note 67, article 1(1), covered only electronic databases. The E.C. Common Position found this distinction unworkable and could not justify differing levels of protection on this basis. See, e.g., Hunsuker, note 33, at p. 6 n. 14 (citing authorities and adding that "today's high speed scanners and optical character recognition software make electronic conversion of nonelectronic databases almost as easy as electronic conversion of electronic databases").

89. See E.C. Directive on Databases, note 52, articles 7(1) (providing initial 15-year term from date of completion), 7(2) (extending protection for an additional 15 years if the database "is made available to the public in whatever manner" before expiration of the initial term), 7(3) (allowing 15-year renewals for "[a]ny substantial change, evaluated qualitatively or quantitatively, to the contents of a database . . . from the accumulation of successive additions, deletions or alterations, which . . . result in . . . a substantial new investment").

90. See id., article 7.

91. See E.C. Directive on Databases, note 52, articles 3 and 5.
92. See E.C. Common Position, note 52, articles 8-9, 16(3).
93. Compare E.C. Directive on Databases, note 52, article 7(2)(a) (defining "extraction" to mean "the permanent or temporary transfer of all or a substantial part of the contents of a database to another medium by any means or in any form") with id., article 5(a).
94. Compare id., articles 7(2)(b) (defining "reutilization" to mean "any form of making available to the public all or a substantial part of the contents of a database by the distribution of copies, by renting, by on-line or other forms of transmission") with id., article 5(b), (d), (e). A database embodied in a hard copy and sold as such remains subject to the first-sale doctrine even under the *sui generis* right, which means that the database maker cannot "control resale of that copy [by the vendee] within the Community." Id., article 7(2) (b). Moreover, public lending of such a copy, say, by a library, "is not an act of extraction or re-utilization." Id., article 2.
95. See note 5 and accompanying text (citing 1996 WIPO documents favoring an international database regime *as proposed by the United States* and international copyright reforms concerning on-line transmissions *as proposed by the European Union*).
96. E.C. Directive on Databases, note 52, article 9(b).
97. See id., articles 9, 9(b).
98. See id., articles 8(2), 9(b).
99. See, e.g., McManis, "Emerging Computer Technology," note 64, at 256 (stating that "[a]ny other substantial extraction from an electronic database [besides illustration for teaching or scientific research] will be infringing, irrespective of whether the extraction is for a commercial purpose, such as market research or private investment decisions, or for a wholly non-commercial purpose, such as religious canvassing, political polling, genealogical research, or pursuit of any . . . hobby or avocation"). McManis contrasts this provision unfavorably with "the exceptions and limitations that safeguard the public interest in copyright law." Id., at p. 54, n. 204, pp. 56-57.
100. See, e.g., Hunsuker, note 33 (stressing fact that article 9(b) speaks of extraction for the *purposes* of illustration for teaching or scientific research, whereas article 6, concerning copyrightable databases, speaks of "the sole purpose of illustration for teaching or scientific research").
101. See note 3 (findings of OTA Report).
102. See generally Reichman, *Collapse of the Patent-Copyright Dichotomy,* note 8, at pp. 488-489 (discussing economic implications and contradictions of such paradoxes).
103. See E.C. Common Position, note 52, article 16(3); E.C. Directive on Databases, note 52, article 16(3).
104. Rosler, note 28, at pp. 138, 140 (stressing tendencies of "[m]onopolists typically [to] charge large premiums for their goods").
105. See, e.g., Rosler, note 28, at pp. 141-43; Reichman, "Legal Hybrids," note 27, at pp. 2496-98.
106. See *Feist Publications, Inc. v. Rural Telephone Service Co.*, 111 S. Ct. 1282 (1991); *Bonito Boats, Inc. v. Thunder Craft Boats, Inc.*, 489 U.S. 141 (1989).
107. See, e.g., Reichman, "Electronic Information Tools," note 15, at pp. 466-67, 472-75.
108. See Robert W. Kastenmeier and Michael J. Remington (1985), "The Semiconductor Chip Protection Act of 1984: A Swamp or Firm Ground?," *Minn. L. Rev.*, 70:417, 438-42 (stating that proponents of new intellectual property laws have the burden to "show . . . that a meritorious public purpose is served by . . . proposed congressional action," and setting out a four-pronged test of public interest that should be met in each case).
109. Simon Chalton (1994), "The Amended Database Directive Proposal: A Commentary and Synopsis," *EIPR,* 16:94, 99 (stressing that national treatment would apply to copyrightable databases, but not to the extraction right). The reciprocity approach, in a more nuanced form, was retained in the final directive. See E.C. Directive on Databases, note 52, article 11.
110. See 17 U.S.C. §§902(a)(1)(A)-(C), 913, 914 (1994); Jay A. Erstling (1989), "The Semiconductor Chip Protection Act and Its Impact on the International Protection of Chip Designs," *Rutgers*

Computer Tech. L. J., 15:303; Charles R. McManis (1988), "International Protection for Semiconductor Chip Designs and the Standard of Judicial Review of Presidential Proclamations Issued Pursuant to the Semiconductor Chip Protection Act of 1984," *Geo. Wash. J. Int'l L. Econ.*, 22:331.

111. Final Act Embodying the Result of the Uruguay Round of Multilateral Negotiations, Marrakesh Agreement Establishing the World Trade Organization, signed at Marrakesh, Morocco, April 5, 1994, Annex 1C, Agreement on Trade-Related Aspects of Intellectual Property Rights (TRIPS Agreement), articles 1(2), (3), 2(1), 3(1), 4, 9(1), 39; McManis, "Emerging Computer Technology," note 64, at pp. 253-62; Paul E. Geller (1995), "Intellectual Property in the Global Marketplace: Impact of TRIPS Dispute Settlement," *Int'l L.*, 29:99, 110.

112. See E.C. Directive on Databases, note 52, article 11 and Preamble, Recital 56.

113. See, e.g., J.H. Reichman (1995), "Universal Minimum Standards of Intellectual Property Protection Under the TRIPS Component of the WTO Agreement,", *Int'l L.*, 29:347-51; Adrian Otten and Hannu Wager (1996), "Compliance with TRIPS: The Emerging World View," *Vand. J. Transnat'l L.*, 29:391.

114. See, e.g., Intellectual Property Committee (USA), Keidanren (Japan), and UNICE (Western Europe) (1988), *Basic Framework of GATT Provisions on Intellectual Property: Statement of the Views of the European, Japanese and United States Business Communities*; see also R. Michael Gadbaw (1989), "Intellectual Property and International Trade: Merger or Marriage of Convenience?," *Vand. J. Transnat'l L.*, 22:223.

115. See, e.g., Hanns Ullrich (1995), "TRIPS: Adequate Protection, Inadequate Trade, Adequate Competition Policy," in *Antitrust: A New International Trade Remedy?*, Pacific Rim Law & Policy Assoc. at pp. 153, 184-207 (John O. Haley and Hiroshi Iyori, eds.); Ralph Oman (1994), "Intellectual Property After the Uruguay Round," *J. Copyright Soc'y*, 42:18 (approving this trend); see generally Reichman, "From Free Riders to Fair Followers," note 59.

116. See, e.g., Morton David Goldberg (1996), "The Digital Agenda in the U.S. and WIPO," paper presented to the Fourth Annual Conference on International Intellectual Property Law and Policy, Fordham University School of Law, April 11-12 (hereinafter Fourth Fordham Conference); Paul Waterschoot (Director, DG XV/E, European Commission) (1996), "Intellectual Property and the Global Information Infrastructure—The E.U. Perspective," paper presented to the Fourth Fordham Conference; *see also* Shira Perlmutter (Associate Register for Policy and International Affairs, U.S. Copyright Office) (1996), "Developments in WIPO: A Status Report on the New Instrument and Protocol," paper presented to the Fourth Fordham Conference. However, the delegations to the Geneva Diplomatic Conference in December 1996 rejected or modified many of these proposals, and a more socially balanced treaty was actually adopted. See WIPO Copyright Treaty, WIPO doc. No. CRNR/DC/89, December 20, 1996, adopted by the Geneva Diplomatic Conference on the same date.

117. See *Committee of Experts on a Possible Protocol to the Berne Convention, Proposals of the European Community and Its Member States*, Geneva, May 22-24, 1996, WIPO doc. BCP/CE/VII/1—INR/CE/VI/1, May 20, 1996; and U.S. White Paper, note 38.

118. See, e.g., U.S. White Paper, note 38, at pp. 114-24; P. Samuelson, "Copyright Grab," *Wired* 4.01:136, 190-91; McManis, note 64, at pp. 68-70 (criticizing this view).

119. See, e.g., U.S. White Paper, note 38, at pp. 230-34; McManis, note 64, at pp. 271-79. There are also privacy reasons behind these measures. Cf. Branscomb, note 3 (who approves of the encryption and information management proposals).

120. See, e.g., Reichman and Samuelson, note 1; McManis, note 64, at p. 73. See also, J.H. Reichman (1995), "The Know-How Gap in the TRIPS Agreement: Why Software Fared Badly, and What Are the Solutions," *Hastings Commun. Ent. L. J. (Commun./Ent.)*, 17:763, 779-84 (citing authorities).

121. See, e.g., McManis, note 64, at pp. 274-76 (stressing proposal to limit removal of electronic "shrink-wrap licenses" as component of U.S. White Paper's overall efforts "to reduce . . . application and scope of fair use doctrine"). For judicial and scholarly opposition to such

licenses, see note 55; see also Charles R. McManis (1993), "Intellectual Property Protection and Reverse Engineering of Computer Programs in the United States and European Community," *High Tech. L. J.,* 8:25, 88-96 (concluding that contracts, or at least shrink-wrap licenses, that prohibit reverse engineering are preempted by federal intellectual property law); Julie E. Cohen (1995), "Reverse Engineering and the Rise of Electronic Vigilantism: Intellectual Property Implications of 'Lock-out' Programs," *So. Cal. L. Rev.,* 68:1091. But see Raymond T. Nimmer (Reporter for the Drafting Committee on Uniform Commercial Code, Article 2B (licenses)), U.C.C. Revision: Information A.S.E. in Contracts (April 15, 1996) (arguing that proposed Art. 2B of U.C.C. should make such licenses presumptively valid); *Pro-CD, Inc. v. Zeidenberg,* 86 F.3d 1447 (7th Cir. 1996) (upholding shrink-wrap license concerning electronic database).
122. McManis, "Emerging Computer Technology," note 64, at p. 67 (citing authorities); Hamilton, note 59, at pp. 628-29.
123. See note 5.
124. See Committee of Experts on a Possible Protocol to the Berne Convention, *Proposal of the United States of America on Sui Generis Protection of Databases,* Geneva, May 22-24, 1996, WIPO doc. BCP/CE/VII/2-INR/CE/VI/2, May 20, 1996 (hereinafter U.S. Proposal on Databases); see also Mark Powell (1996), "The European Union's Database Directive: An International Antidote to the Side-Effects of *Feist,*" paper presented to the Fourth Fordham Conference. Powell notes that the E.C. Directive will be incorporated into the laws of Norway, Iceland, and Liechtenstein under existing trade agreements with the E.U.; that "the Commission will encourage Central and Eastern European countries to adopt similar legislation" in their Association Agreements; that the E.U.-Turkey Customs Union Decision explicitly obliged Turkey to align its legislation on databases with the Directive; and that its reciprocity clause "will be used by the Commission as a bargaining chip" in dealing with third countries.
125. Powell, id.
126. See, e.g., Powell, note 124, at pp. 2-3 (objecting that "it is questionable whether an international instrument should be founded on a legal measure with no proven track record and which contains such novel legal concepts . . . especially since . . . [n]either database makers nor users were satisfied with the compromise reached in the Directive"); see also Pamela Samuelson (1994), "The N.I.I. Intellectual Property Report," *Communications of the ACM,* 37:17. (finding it "peculiar that the WIPO experts should even consider recommending a treaty on database protection when the idea for such a law is so new and untested").
127. See, e.g., Powell, note 124, at p. 16 (stating that the "economic case for the creation of a right to prevent extraction and/or re-utilization of non-original contents by users has never been satisfactorily explained"); see also id., at p. 55 (stressing that an "international treaty, . . . is trickier to modify" than domestic models and suggesting that the Diplomatic Conference should confine itself to a "set of general principles" for legal protection of databases that would be reviewed within a fixed period of time).
128. See, e.g., Jens L. Gaster (Principal Administrator, DG XV-E-4, European Commission) (1996), "The New E.U. Directive Concerning the Legal Protection of Data Bases," paper presented to the Fourth Fordham Conference (conceding that "the *sui generis* right was considerably strengthened during the legislative process" and that attacks on the right to extract even insubstantial parts of a protected database were barely repelled).
129. TRIPS Agreement, note 111, article 71; see also id., articles 68-69. While parties to the Berne Convention remain free to adopt higher copyright standards among themselves (see Berne Convention, note 63, article 20), these arrangements would not become binding on other Berne members in the absence of a unanimous decision. See id., article 27(3). Unless such standards were incorporated into the TRIPS Agreement, parties to a special arrangement under Berne (or related to Berne) would run some risk of having to extend the higher standards to nonsignatory members of the WTO, under the most-favored-nation clause of the TRIPS Agreement. See TRIPS Agreement, note 111, article 4. While applications of Article 4 remain inherently

uncertain, and this outcome would depend on the interpretation of various provisions in both the TRIPS Agreement and prior international agreements (see notes 63 and 111 and accompanying text), the goal is clearly to develop "a model in the search for a global solution regarding the protection of databases which is presently discussed at WIPO." Gaster, note 128.
130. See TRIPS Agreement, note 111, articles 1(3), 2(2), 3(1), 35-38; J.H. Reichman, "Universal Minimum Standards of Intellectual Property Protection Under the TRIPS Agreement," *Int'l. L.,* 29: 374-375.
131. See U.S. Congress, House of Representatives, H.R. 3531, 104th Congress, 2d Session, May 23, 1996, sec. 1 (short-titled "Database Investment and Intellectual Property Antipiracy Act of 1996") (hereinafter H.R. 3531). This bill was introduced by Carlos Moorhead, chairman of the House Subcommittee on Courts and Intellectual Property, and referred to the Committee on the Judiciary.
132. See id., sections 2, 3(a).
133. See id., sections 2, 3(a), (b), 6.
134. U.S. Constitution, Article I, Section 8, clause 8.
135. See, e.g., Reichman and Samuelson, note 1 at pp. 102-09, 132-36. Because the European Community has codified a doctrine of "thin" copyright protection of computer programs, it expressly subordinated the protection of databases to that policy (see E.C. Directive on Databases, note 52, article 2(a)); in contrast, because the U.S. federal courts have resisted a policy of "thick" or strong copyright protection for computer programs, there is reason to fear that the *sui generis* database law may be used to overturn these precedents.
136. H.R. 3531, note 131, section 4(a)(1).
137. This provision is thus consonant with several other key provisions that greatly strengthen the scope of protection in general. See *infra* text accompanying notes 138-155.
138. See H.R. 3531, note 131, article 6(b); notes 89-90 and accompanying text.
139. See E.C. Directive on Databases, note 52, article 8(1) at p. 26. Gaster, note 128, at pp. 9, 11 (indicating that protection of the right to extract—but not to reuse—an insubstantial component was an integral part of the compromise that led to otherwise strengthened protection).
140. See H.R. 3531, note 131, section 4(a)(2).
141. Id.; E.C. Directive on Databases, note 52, article 7(g).
142. See H.R. 3531, note 131, sections 4(a)(1), (2), 5(a).
143. See id., section 4(a)(2), 4(b). This restriction covers markets in which the database owner has a demonstrable interest or expectation in licensing or otherwise reusing the database, as well as markets in which customers might reasonably be expected to become customers for the database.
144. Id., section 4(b) (4).
145. See E.C. Directive on Databases, note 52, article 15 (expressly voiding contractual provisions to this effect).
146. See Powell, note 124, at pp. 40-43. See also U.S. Proposal on Databases, note 124, article 7(2) ("No Contracting Party shall impair the ability to vary by contract the rights and exceptions to rights set forth herein").
147. See H.R. 3531, note 131, sections 4(a) (2), 4(b) (4).
148. See text accompanying notes 91-92 and Box 5.4; Reichman and Samuelson, supra note 1, pp. 103-109.
149. See, e.g., H.R. 3531, note 131, section 5(B), which states: "Nothing in this Act shall in any way restrict any person from independently collecting, assembling or compiling works, data or materials from sources other than a database subject to this Act."
150. See H.R. 3531, note 131, sections 4, 5, 6.
151. See E.C. Directive on Databases, note 52, article 9(b); see also id., articles 9(a) (allowing extraction for private purposes from nonelectronic databases), and 9(c) (allowing extraction and reuse for purposes of "public security or an administrative or judicial procedure").

152. See H.R. 3531, note 131, section 5; U.S. Proposal on Databases, note 124, article 5. The U.S. Proposal to WIPO appears less watertight, because it does permit contracting parties, "in their domestic legislation, [to] provide for exceptions to or limitations on the rights," so long as such exceptions or limitations "do not unreasonably conflict with a normal exploitation . . . and do not unreasonably prejudice the legitimate interests of the rightholder." Id., article 5.3. Because the U.S. Proposal links this exception to the notion of a "substantial" taking for purposes of infringement and also to the express notion that use of preexisting protected matter is not independent creation (see id., articles 3.1, 3.2), the drafters clearly aim to forbid any exceptions that permit extraction or use of a substantial part of the database for any purpose. At the same time, the U.S. Proposal to WIPO is more amenable to local variants that expand upon uses of insubstantial components than is H.R. 3531, and to this extent it moves further towards the E.C. Directive.
153. See H.R. 3531, note 131, section 3(c), which provides: "Except for a database made by a governmental entity, any database otherwise subject to this Act, is not excluded herefrom because its contents have been obtained from a governmental entity."
154. See H.R. 3531, note 131.
155. See Executive Office of the President, Office of Management and Budget, "Implementing the Information Dissemination Provisions of the Paperwork Reduction Act of 1995," Memorandum by Alice M. Rivlin, September 22, 1995 (cautioning agencies that use the services of private contractors not to impose, or permit the intermediary to impose, restrictions that interfere with the agencies' own dissemination responsibilities; and reiterating "the basic standard that agencies shall not charge use fees for government information which exceed the cost of dissemination").
156. See H.R. 3531, note 131, sections 4, 5(a)(b); text accompanying notes 140-147 (stressing built-in restrictions on claiming use of an insubstantial part in practice).
157. Cf. E.C. Directive on Databases, note 52, article 16(3) (requiring E.C. Commission to report, at three-year intervals, concerning these issues and the need to establish "non-voluntary licensing arrangements").
158. See, e.g., H.R. 3531, note 131, section 10 ("Circumvention of Database Protection Systems"); U.S. Proposal on Databases, note 124, article 8 (Prohibition of Protection—Defeating Devices).
159. See, e.g., H.R. 3531, note 131, section 11 ("Integrity of Database Management Information").
160. See, e.g., Branscomb, note 3, at pp. 175-77.
161. See text accompanying note 15.
162. See H.R. 3531, note 131, sections 10, 11.
163. See further *infra* text accompanying notes 137-138.
164. See H.R. 3531, note 131, sections 8, 13.
165. See David and Foray, note 35.
166. See Perlmutter, note 116; Morton David Goldberg (1996), "The Digital Agenda in the U.S. and WIPO," paper presented to the Fourth Fordham Conference.
167. See note 124.
168. Cf., e.g., Dirk J.G. Visser (1996), "Copyright in Cyberspace—National Dutch Report," paper presented to the International Association for Literary and Artistic Property (ALAI) Study Days, Amsterdam, June 4-8, at p. 12, quoting the Dutch Federation of Organizations in the Library, Information and Documentation Fields' (FOBID) recent complaint to the Minister of Justice:

> To its unpleasant surprise FOBID has found that in the [E.C.] Green Paper [see note 4] little or no attention is paid to the statutory limitations on copyright, such as library privileges and rules on educational, scientific and private use. Many existing limitations are technology dependent. It has to be examined whether and to what extent these limitations should be maintained or adapted in the digital environment.

To this and similar complaints, Visser's report makes the following reply:

[M]any limitations are the result of successful "lobbying." Intermediaries and users applying for specific limitations must realize that right owners, who will oppose any limitation as a matter of principle, are generally very well represented at the (national and international) legislative level. Thus, the extent to which copyright limitations will be preserved or extended in the digital environment will eventually be determined by the ability of intermediaries and users to have their voice heard on the political level.

Visser, at p. 13 (quoting P.B. Hugenholtz and D.J.G. Visser (1995), *Copyright Problems of Electronic Document Delivery: A Comparative Analysis*, Report to the Commission of the European Communities (DG XIII), Brussels/Luxembourg, p. 62.) See also Robert J. Hart, "Intellectual Property and the Global Information Infrastructure—The Perspective in Japan, Australia, and Canada," paper presented to Fordham's Fourth Conference (stressing that only Australia's proposals concerning regulation of national information infrastructures have so far reflected concerns for "fair use" and related exceptions).

169. See Berne Convention, note 63, article 20 (authorizing member states "to enter into special agreements among themselves . . . [that] grant to authors more extensive rights than those granted by the Convention").

170. The next step would be to use the Council for TRIPS and relevant provisions of the TRIPS Agreement as a springboard for binding non-Berne countries, and especially developing countries, to the same standards. See, e.g., J.H. Reichman (1995), "Universal Minimum Standards," note 130, at pp. 345, 383-85; J.H. Reichman, "From Free Riders to Fair Followers," note 59 (urging developing countries to resist these pressures).

171. See, e.g., E.C. Commission, Green Paper on Copyright and Related Rights in the Information Society (July 19, 1995); Paul Waterschoot, "An Overview of Recent Developments in Intellectual Property in the European Union," paper presented to Fourth Fordham Conference, note 116.

172. See, e.g., McManis, note 64, at p. 211.

173. John Perry Barlow (1994), "The Economy of Ideas: A Framework for Rethinking Patents and Copyrights in the Digital Age (Everything You Know About Intellectual Property is Wrong)," *Wired 2.03*, at p. 84 (March).

174. See, e.g., Paul Edward Geller (1996), "Conflicts of Law in Cyberspace: Rethinking International Copyright in a Digitally Networked World," *Colum.-VLA J. L.*, 20:571; I. Trotter Hardy (1994), "The Proper Legal Regime for 'Cyberspace,'" *U. Pitt. L. Rev.*, 55:993; Jane C. Ginsburg (1995) "Global Use/Territorial Rights: Private International Law Questions of the Global Information Infrastructure," *J. Copyright Soc. USA*, 42:318.

175. See, e.g., Reichman, "Universal Minimum Standards," note 130, at pp. 347-51, 365-73.

176. For the possibility that some developing countries might postpone the obligation to implement these standards beyond the 11-year transitional period specified by the TRIPS Agreement, on grounds of hardship, see Reichman, "Universal Minimum Standards," at p. 353, note 130.

177. See, e.g., Reichman, "From Free Riders to Fair Followers," note 59 (arguing that if developing countries interpret "grey areas" of TRIPS Agreement in a procompetitive manner, it could benefit consumers, users, and second comers everywhere).

178. See, e.g., Reichman, "Know-How Gap in TRIPS," note 120, at pp. 779-84 (uncertain scope of copyright protection for computer programs); Reichman, "Legal Hybrids," note 27, at pp. 2459-64, 2488-90 (uncertain status of industrial designs under diverse regimes, including copyright law).

179. McManis, "Emerging Computer Technology," note 64, at p. 284; see also Kurtz, note 32, at pp. 120, 121, 124 (stressing risk that a chronic state of overprotection could "choke off opportunities for academic research and educational uses of intellectual property").

180. Rosler, note 28, at pp. 141-42.

181. The OMB circular *supra*, note 155, suggests that pressures to privatize the distribution of government-funded data had already become a problem before the advent of the proposed database law in the United States.
182. See also Kurtz, note 32, at p. 121 (stating that overprotective legal monopolies "can . . . be as stifling to creation as underprotection . . . [and they] can choke off opportunities for academic research and educational uses of intellectual property"). Even though the sole-source provider may not wish to price itself out of the market, this will be cold comfort to "those who cannot afford to pay, and [could thus] lead to a society of information haves and have nots."
183. See David and Foray, note 35.
184. See Reichman and Samuelson, note 1, at pp. 113-30, 137-63.
185. See text accompanying notes 73-74.
186. See, e.g., Ginsburg, "Information After *Feist*," note 23, at pp. 384-387.
187. The ambivalence of the final E.C. Directive in this regard is explained in part by the fact that no serious fair-use provisions had previously been developed in the presence of the compulsory license that the Council of Ministers deleted at the last moment, and in part by the growing disinclination of both the European Union's and the United States' intellectual property authorities to recognize fair use in the digital environment.
188. Goldstein, *Celestial Jukebox*, note 21, at pp. 220-22. Gordon, "Fair Use as Market Failure," note 22.
189. Samuelson et al., "Manifesto," note 38, at pp. 2385-86 (discussing limits of legal protection for single features of computer programs).
190. See, e.g., OMB Circular A-130, U.S. Government Printing Office. For the problem of leakage as a limit on price discrimination, see Chapter 4.
191. See E.C. Directive on Databases, note 52, article 6(2).
192. For example, "NASA and Orbital Sciences . . . reached an agreement regarding the Sea WiFS mission for ocean color data," a private endeavor "for which NASA provided upfront money for a data purchase . . . [so that] the companies . . . [could] get financial backing." Under this agreement, "the company had exclusive rights to exploit the data for a number of days, after which the data went to NASA for scientific purposes." Letter from Joanne Gabrynowicz, July 25, 1996. Reportedly, this agreement was possible owing to the perishability of ocean color data for commercial purposes. However, for some scientific disciplines, which depend on real-time observations, delay as a form of product differentiation is not feasible. This, in turn, suggests the importance of legal measures that permit providers and users to adjust the concept of fair use (or fair and reasonable terms) to the needs of different categories of users.
193. See, e.g., Reichman and Samuelson, note 1, at pp. 158-61. For the general importance of such a clause in the post-TRIPS environment, especially with regard to transfer of technology agreements, see Reichman, "From Free Riders to Fair Followers," note 59.
194. For example, licenses "issued pursuant to federal law for private remote sensing systems require that system operators make their commercially obsolete data available to the National Data Archive on 'reasonable terms and conditions.' The government does not require that they [be] give[n]. . . the data, nor does it set the criteria by which the decision is made. But if, and when, a company decides to purge data, it triggers the requirement, [which] . . . amounts to the government having the right of first refusal." Letter from Joanne Gabrynowicz, July 25, 1996.
195. See, e.g., E.C. Directive on Databases, note 52, articles 7(5), 8(2) (forbidding lawful user of database to perform acts that "conflict with normal exploitation" or that "unreasonably prejudice [the maker's] legitimate interests"; see also H.R. 3531, note 131, sections 4(a), (b), 5.
196. See Reichman and Samuelson, note 1, at pp. 145-51.
197. Goldstein, *Celestial Jukebox*, quoted note 21; see also Ginsburg, "Information After *Feist*," note 23, at pp. 386-87 (deeming compulsory licenses indispensable under a noncopyright protection scheme).

198. Cf. TRIPS Agreement, note 111, article 31(b) (allowing right of states to impose compulsory licenses on foreign patentees only if, prior to the grant, "the proposed user has made efforts to obtain authorization from the right holder on reasonable commercial terms and conditions").
199. See, e.g., Hunsuker, note 33.

APPENDIXES

APPENDIX
A

Abbreviations and Acronyms

ADDS	African Data Dissemination Service
ADS	Astrophysics Data System
ASCII	American Standard Code for Information Interchange
ATM	asynchronous transfer mode
AURA	Association of Universities for Research in Astronomy
AVHRR	Advanced Very High Resolution Radiometer
BASIS	Battelle Automatic Search Information System
CDC	Centers for Disease Control and Prevention
CDF	Collider Detector at Fermilab
CDIAC	Carbon Dioxide Information Analysis Center
CD-ROM	Compact Disk-Read Only Memory
CEC	Commission of the European Communities
CEOS	Committee on Earth Observation Satellites
CGIAR	Consultative Group on International Agricultural Research
CIESIN	Consortium for International Earth Sciences Information Network
CINDAS	Center for Information and Numerical Data Analysis and Synthesis
COBIOTECH	Committee for Biotechnology (ICSU)
CODATA	Committee on Data for Science and Technology (ICSU)
COSTED	Committee on Science and Technology in Developing Countries

DIS	data indexing system; data and information system
DMC	Data Management Center (IRIS)
DNA	deoxyribonucleic acid
DOC	Department of Commerce
DOD	Department of Defense
DOE	Department of Energy
DOI	Department of the Interior
DVD	digital versatile disk
ENDF/B	Evaluated Nuclear Data File/B
ENSDF	Evaluated Nuclear Structure Data File
EOS	Earth Observing System
EOSAT	Earth Observation Satellite Company
EOSDIS	Earth Observing System Data and Information System
EPA	Environmental Protection Agency
EROS	Earth Resources Observation Systems
ESA	European Space Agency
FAME	Fatty Acid Methyl Ester
FAO	Food and Agriculture Organization
FDSN	Federation of Digital Seismic Networks
FITS	Flexible image transport system
FTP	file transfer protocol
GIS	Geographic Information System
GMT	Greenwich mean time
HD-ROM	high density-read only memory
HEP	high-energy physics
HIPPI	high-performance parallel interface
HST	Hubble Space Telescope
HTML	HyperText Mark-up Language
HTTP	HyperText Transfer Protocol
IAEA	International Atomic Energy Agency
ICSU	International Council of Scientific Unions
IEEE	Institute for Electrical and Electronics Engineers
IETF	Internet Engineering Task Force
IFS	International Foundation for Science
IGBP	International Geosphere-Biosphere Programme
IITF	Information Infrastructure Task Force
IMPACT	Information Market Policy Action Program
IPng	Internet Protocol next generation

APPENDIX A 193

IPR	intellectual property right
IRAF	Image Reduction and Analysis Facility
IREX	International Research and Exchange Board
IRIS	Incorporated Research Institutions for Seismology
ISDN	integrated services digital network
ISO	International Organization for Standardization
LAN	local area network
MBone	multicast broadcast backbone network
Mbps	megabits per second
MEMS	micro-electromechanical systems
MIAC	Metals Information Analysis Center
MSS	multispectral scanner
NAS	National Academy of Sciences
NASA	National Aeronautics and Space Administration
NCBI	National Center for Biotechnology Information
NCDC	National Climatic Data Center
NGDC	National Geophysical Data Center
NGO	nongovernmental organization
NIH	National Institutes of Health
NII	national information infrastructure
NIST	National Institute of Standards and Technology
NLM	National Library of Medicine
NNDC	National Nuclear Data Center
NOAA	National Oceanic and Atmospheric Administration
NODC	National Oceanographic Data Center
NRC	National Research Council
NSF	National Science Foundation
NUDAT	Nuclear database (part of ENSDF)
OAS	Organization of American States
OECD	Organisation for Economic Co-operation and Development
OMB	Office of Management and Budget
OTA	Office of Technology Assessment
PC	personal computer
PDF	PostScript Data File
POSIX	Portable Operating System Interface for Computer Environments (an IEEE standard)
POTS	"plain old" telephone service
PTO	Patent and Trademark Office

PTT	Post, Telephone, and Telegraph ministry
RNA	ribonucleic acid
SGML	Standard Generalized Mark-up Language
SIMBAD	set of identifications, measurements, and bibliography for astronomical data
SMART	System for the Mechanical Analysis and Retrieval of Text
SMTP	Simple Message Transfer Protocol
SNMP	Simple Network Management Protocol
SOTER	Soils and Terrain Digital Database
SPECINFO	a database from Chemical Concepts GmbH containing digitized nuclear magnetic resonance data and infrared spectra for chemical compounds
STAIRS	Storage and Information Retrieval System
ST-ECF	Space Telescope European Coordinating Facility
STN	Science and Technology Network International
STScI	Space Telescope Science Institute
TCP/IP	Transmission Control Protocol/Internet Protocol
TM	Thematic Mapper
TRIPS	Trade-Related Aspects of Intellectual Property Rights, Agreement on
TWAS	Third World Academy of Sciences
UC	University of California
UNCED	United Nations Commission on Economic Development
UNCTAD	United Nations Commission on Trade and Development
UNDP	United Nations Development Programme
UNEP	United Nations Environment Programme
UNESCO	United Nations Educational, Scientific, and Cultural Organization
UNIDO	United Nations Industrial Development Organization
URL	uniform resource locator
URN	uniform resource name
USAID	United States Agency for International Development
USGS	United States Geological Survey
USNC-CODATA	United States National Committee for CODATA
VITA	Volunteers in Technical Assistance
VHS	video home system

WDC	World Data Center
WHO	World Health Organization
WIPO	World Intellectual Property Organization
WMO	World Meteorological Organization
WTO	World Trade Organization
WWW	World Wide Web

APPENDIX
B

Glossary

Access. The process of obtaining data from a storage device or system.
Analog data. Data in the form of continuously variable physical quantities. [C]
Animation. The use of computer graphics (or sometimes cinematography) to synthesize drawings or other objects so as to depict motion.
Archival database. A database containing data values and other information retained over a period of time and represented as an accurate reflection of the contents at a specified time. [C]
Archive. An organized and managed collection of information (in any form) that is protected to ensure its integrity as an authoritative source for the information stored in it.
Artificial intelligence. The capability of a device to perform functions that are normally associated with human intelligence, such as reasoning, learning, and self-improvement. [C]
Asynchronous transfer mode. The transmission of information in irregular sections, with the time interval of each transmission varying and each section being identified by a start and stop signal. [C]
Authentication. The process by which a prospective computer user's identity is verified by a card, token, or biometric device before system access is allowed. [C]

NOTE: The sources of the definitions appearing in this glossary are indicated by the symbols in boldface square brackets. These sources are spelled out in detail below. Where no source is shown, the definition was crafted explicitly for this glossary by the committee.

Backbone. (1) A set of nodes and their interconnecting links providing the primary data path across a network . . . . A backbone may be configured as a bus or as a ring; (2) in a wide area network, a high-speed link to which nodes or data switching exchanges are connected. **[IBM]**

Bookmark method. Refers to a system designed to leave a marker in an application or presentation to which the system will return when the program is next executed.

Bridge. A functional unit that interconnects two local area networks that use the same logical link control protocol but may use different access control protocols. **[IBM]**

Broadband transmission. In communications, pertaining to transmission facilities whose bandwidth is greater than that available on voice grade circuits and therefore capable of high-speed data transmission. **[L&S]**

Browser. A program used to scan and search electronic files.

Bulletin board. A set of files, stored on a computer, that may be accessed by any user with a terminal and that is controlled by a systems operator, who organizes the files into topical areas. Files are uploaded and downloaded by users, and most bulletin boards allow users to exchange electronic mail. Computer suppliers often maintain a bulletin board to allow users to have access to information system updates and to report on problems encountered. **[D]**

Cache. (1) A special-purpose buffer storage, smaller and faster than main storage, used to hold a copy of instructions and data obtained from main storage and likely to be needed next by the processor; (2) a buffer storage that contains frequently accessed instructions and data; it is used to reduce access time; (3) an optional part of the directory database in network nodes where frequently used directory information may be stored to speed directory searches. **[IBM]**

Client/server. A network system designed to share substantial resources that cannot be provided on every user's machine. The requesting program (client) sends a request to a program at another site (server) and awaits a response.

Compilation. (1) With respect to computer programs, the result of the implementation of a compiler program used to make a high-language program operable in a computer; (2) the result of aggregating several documents or files. **[C]**

Cyberspace. The invisible, intangible world of electronic information and processes stored at multiple interconnected sites, with controlled access and manifold possibilities for interaction.

Data. Scientific or technical measurements, values calculated therefrom, and observations or facts that can be represented by numbers, tables, graphs, models, text, or symbols and that are used as a basis for reasoning or further calculation.**[C]**

Data collection. The process of identifying and accumulating data and/or data sets pertaining to a common subject, application, or theme. [C]

Data exchange. (1) The transfer of data between or among more than one computer program or system; (2) for people or organizations, the sharing of data between or among people and/or organizations.

Data integrity. The property/quality of a collection of data being complete, consistent, and accurate. [C]

Data management/data administration. (1) The function of controlling the acquisition, analysis, storage, retrieval, and distribution of data; (2) in an operating system, the programs that provide access to data, perform or monitor organization and storage of data, and control input/output devices. [C]

Data processing. The systematic performance of operations on data, such as handling, merging, sorting, and computing. [C]

Data transfer. Transmission, copying, or movement of digitized data from one place to another, such as from database to database or computer to computer. [C].

Data validation. A process used to determine whether data are accurate, complete, and/or reasonable. [C]

Database (also data set). A collection of interrelated data, often with controlled redundancy, organized according to a schema to serve one or more applications. The data are stored so that they can often be used by different programs with little or no restructuring or reorganization of the data. A systematic protocol is used to add new data or modify and retrieve existing data. [C]

Database management. The activity associated with organizing, storing, and providing access to a computerized database. It usually includes responsibility for ensuring the integrity of the database. [C]

DECnet. A family of software products supporting the general interconnection of computer systems of Digital Equipment Corporation; also used to refer to a network that uses this software.

Decryption. The process that reverses encryption, turning a file into plain text.

Derogation. An act partly repealing something, or lessening it in value, force, authority, rank, honor, or the like.

Desktop publishing. The local production of high-quality, camera-ready copy by an individual or individuals by means of sophisticated word-processing, graphics, and other software packages and laser printers.

Digital signature. A means by which the recipient of a message can validate that it came from the person who claims to have sent it. A sender can create a signature for a message and encrypt it with his or her private key. The enciphered signature can be decrypted only by the sender's corresponding public key known to the recipient. A signature encrypted by any other private key, corresponding to another individual or source, cannot be decrypted by the same public key.

Electronic mail (also **e-mail**). A method of sending messages from a user on one computer to the computer of another user. In its simplest form, electronic mail on a local area network operates by storing all mail messages on files on one machine designated as the post office. Mail does not move around; users access the post office when they want to "collect" their mail and read it. On larger systems, post offices are interconnected and can exchange mail, so that users on different LANs can also communicate. **[IBM]**

Electronic publishing. (1) Submission of text, graphics, and images in electronic form to a publisher who will work directly on these files with various computer-assisted techniques to ready the material for publication; (2) submission of articles, data tables, graphs, etc., to an electronic network where they may be browsed, annotated, critiqued, or downloaded.

Encryption. A process that transforms data/information, using a key or keyed process, so that only the recipient who has the same key or appropriate keyed process can read it. Traditional encryption uses secret keys that must be exchanged separately among senders and recipients. The same key used for encryption of a message is used for decryption. See also **public-key encryption**.

E-print archive. A collective file of full-text electronic journals, available on the Internet, selections from which can be readily downloaded by individual scientists.

Evaluated data. Data that have been subjected to evaluation and can be regarded as scientifically correct within the limits specified.

Evaluation. The process of establishing the accuracy and integrity of measured data. Evaluation involves examination and appraisal of the data presented, assessment of experimental techniques and associated errors, consistency checks for allowed values and units, comparison with other experimental or theoretical values, reanalysis and recalculation of derived quantities as required, selection of best values, and assignment of probable error or reliability. **[C]**

Experimental science. One founded on tests, acts, or operations carried out under controlled conditions and intended to discover something unknown or to test a principle or hypothesis.

Extraction. The permanent or temporary transfer of all or a substantial part of the contents of a database to another medium by any means or in any form. **[EC]**

Full and open exchange of data. Occurs when data and information are made available with as few restrictions as possible, on a nondiscriminatory basis, for no more than the cost of reproduction and distribution.

Functional dependence. In databases, an indication of the interrelationships of attributes in a relation. Attribute A of a relation R is functionally dependent on attribute B of relation R if, at every instant of time, each value of B has no more than one value of A associated with it. **[L&S]**

Gigabyte. 10^9 bytes, or 8×10^9 bits.

Home page. A file on the Internet that contains information provided by the sponsor of the home page and links to other files (or processes) of the sponsor or to other locations on the Internet.

Hyperlink. The text, number, symbol, or code providing linkage to related items in an electronic file or a network of electronic files.

Hypermedia. A system of presenting information in discrete units, or nodes, that are connected by links. The information may be presented using a variety of media such as text, graphics, audio, video, animation, image, or executable documentation. **[IBM]**

Hypertext. Text coded with internal cross-references allowing users to link to other related material posted on the network. **[C]**

Image. A visually interpretable representation of objects, text, or graphics as displayed, plotted, or printed, that is not semantically coded.

Information. Data that have become meaningful as a result of collection, processing, organization, and interpretation, in light of some hypothesis.

Informatics (especially in continental Europe). The study of the quantitative aspects of information in any form, including patterns of language use, the production, dissemination, and use of recorded information, and the quantitative aspects of science as a discipline or economic activity. **[E]**

Interactive information system. In computing, a system allowing continuous dialog or two-way communication between a user and the computer. **[L&S]**

Internet. A worldwide network of semiautonomous networks that share a common protocol for internetwork communications and a common form of naming of user and server nodes.

Intranet. A dedicated computer network (either a local-or wide-area network) within a corporation or other private institution intended to serve its own needs for data exchange, electronic mail, bulletin boards, etc. with reliability, performance, and security beyond that provided by the public Internet.

"Knowbot." An information robot, a sophisticated browser program for scanning networks of files and employing more elaborate detection schemes than simple Boolean filters.

Knowledge. An organized body of information that represents a description and understanding of concepts, properties, and relationships as viewed by an authoritative community capable of using it.

LaTeX, TeX. Proprietary word-processing programs especially designed for technical writing.

Link. A physical or logical connection between two data items or records.

List server. A mechanism that provides a flow of "issued" information to a routing list maintained by the system.

Megabyte. 10^6 bytes, or 8×10^6 bits.

Metadata. Data about data; consists of descriptors of data in a database to provide systematic information for users, application programs, and database management software. Metadata may be manipulated and searched and may themselves be organized in a database. [C]

Mirror site. A proxy server acting as a cache containing duplicated data from a central database so as to facilitate more economic and reliable access by local users.

Moore's Law. Computer power (speed) will double every 18 months, for an effective reduction in price by half.

Multicasting. The transmission of data from one place to many specific ones, as opposed to broadcasting (which transmits everywhere within a range). An example is the transmission of sound and video over the Internet. [D]

Multimedia. Any system that combines computer sound and video; may incorporate an interactive facility.

Natural sciences. Branches of science such as biology, physics, and chemistry that deal with objectively measurable objects or processes observable in nature, as distinguished from the abstract sciences such as mathematics or philosophy.

Navigation. Following a path within a file or among files to locate desired information.

Near real time. Approximate simultaneity in time; without delay, compression, or expansion in the time variable.

Neural network. A system of multiple parallel structures, containing a highly interconnected set of simple elements, exploiting aspects of their collective behavior and operating according to a strict, deterministic set of rules for proceeding from one step to the next.

Normalize. To reduce data, data structure, or relations among data to a standard form. In the case of numeric data, normalization may involve presenting a data value in terms of its relation to a reference value, e.g., strength at test temperature as a percent of strength at room temperature.

Object-oriented database. A database of objects (entities including data, queues, data structures, procedures, modules, etc.) using the principles of object-oriented programming and the concept of inheritance of methods by new classes of objects.

Observational science. A science founded on observation rather than experimentation.

On-line. Pertaining to an interactive process by which a user and host computer converse.

Overlay. (1) The superposition of two graphical representations for purposes of comparison; (2) one of several segments of a computer program that, during execution, occupy the same area of main [or virtual] storage, one segment at a time. [C]

Packet. In data communication, a sequence of binary digits, including data and control signals, that is transmitted and switched as a composite whole. The data, control signals, and possibly error control information are arranged in a specific format. [C]

Parallel communication. A form of communication that simultaneously transmits on multiple communication paths controlled as a unit.

Portability. Capability for transporting of data or programs between computers or computer systems without any required modification thereof. [C]

PostScript. A proprietary (Adobe Systems) page definition language used in desktop publishing and capable of handling full text, mathematics, and graphics.

Primary data *see* **raw data**

Privatization. Transfer from government management and control to management and control by a private company or other nongovernmental organization.

Protocol. In a computer system, a formalized set of conventions, including coding, prescribed responses to received signals and/or coded messages, wiring patterns, etc., necessary to establish and maintain communication between two devices or systems.

Public good. A product or service typically provided by government that is nondepletable and produces benefits from which others cannot be excluded and which cannot easily be constrained only to those who pay.

Public-key encryption. A method of encryption/decryption in which the enciphering and deciphering keys are different and not computable from each other. A person wishing to receive a secret message can publicly distribute the enciphering key, which, when used by the originator of a message, produces an enciphered message that only the secret deciphering key kept by the recipient can decipher. The first and still most widely used public-key algorithm is the RSA algorithm; the public key is derived from the product of two very large prime numbers, and the strength of protection derives from the size of the prime numbers and the resulting computational difficulty in decomposing the product in those primes.

Raw data. Data as originally recorded and that have not been combined, modified, interpreted, or adjusted in any way.

Real time. In computing, pertaining to operations that are performed in conjunction with some external process or user and that are required to meet the time constraints imposed by that process or user, e.g., control of an airplane guidance system or an on-line information service. [L&S]

Relational database. A database composed of flat files, each embodying different facets of information about the data, organized by linkages representing one or more relations; the resulting data structure allows the efficient use of many types of data during queries and can be manipulated with great flexibility using a formal relational algebra to derive standard (normalized) forms and to establish equivalencies.

Remote access. Capability for communicating with a data processing facility through a data link, usually at a different geographical location.

Remote sensing. Gathering and recording of data through a system of one or several sensors capable of obtaining data from afar and transmitting the values to a central computer or other display, recording, or storage device.

Retrieval. The location and transfer of selected data or information from a source (long-time memory or another information system) to local computer memory for subsequent processing, display, or downloading to another device.

Router. A computer that determines the path of network traffic flow. The path selection is made from several paths based on information obtained from specific protocols, algorithms that attempt to identify the shortest or best path, and other criteria such as metrics or protocol-specific destination addresses. **[IBM]**

Search engine. A device or program for scanning a file to match data patterns for purposes of retrieval.

Secondary user. A user outside the community by or for which the collection of data was initiated.

Serial communication. Transmission of a sequence of signals on a single communication path.

Server. A computer on a network that performs functions (e.g., computation, storage, and retrieval) for other computers, terminals, and peripheral devices on the network.

Shareware. Software distributed publicly by individuals who have developed a useful program and wish to share it informally (as contrasted to commercial marketing) with others. **[adapted from C]**

Spectral data. Data measured for specific wavelengths of radiation related to the phenomenon being observed. The data may be a continuous function of wavelength value or may indicate only values at discrete wavelengths.

Sui generis. Of its own kind; unique.

Terabyte. 10^{12} bytes, or 8×10^{12} bits.

User friendly. Characteristic of a system that inexperienced or untrained users feel comfortable working on, and find helpful and not intimidating. Generally, use of such a system does not depend heavily on knowing or remembering procedural details; rather, the system provides hints, guidance, and choices among alternatives at every stage of the interaction.

Validated data. Data values or data sets that have been shown to be generated according to standard test methods and practices or other measures of quality, reliability, and precision. **[C]**

Verified data. Data that have been shown to be accurately transcribed or transformed from other representations.

Virtual reality. A computer-generated simulation of reality (or imagined reality) with which users can interact using specialized peripherals such as data gloves and head-mounted computer graphic displays. **[L&S]**

Workstation. (1) A term traditionally used to describe a high-end, high-performance assembly of equipment, including extensive computer graphics capability, as an interactive console for performing a specialized function such as computer-aided design, computer animation, or complex data analysis. Typically, the workstation includes enough computing power and data storage capacity to operate free-standing from any other computer, although workstations are often networked to larger systems for data archives, communication with other workstations, etc. As the technology of personal computers has improved, they have taken on functions previously handled on workstations. (2) Any assembly of computer equipment designed to support the full range of work of an individual.

World Wide Web. A tool, based on the hypertext system, for the organization and representation of on-line information available over the Internet. To access the Web, a browser such as Netscape or Mosaic is required. The user must know the address of the home page of the institution whose information is to be accessed. The WWW provides the simplest means of navigating the Internet, giving access to text, images, sound, and video. **[D]**

Sources of Definitions

[C] J.H. Westbrook and W. Grattidge, eds. (1991), "A Glossary of Terms Relating to Data, Data Capture, Data Manipulation, and Databases," *CODATA Bulletin* 23, Nos. 1 & 2, 196 pp.

[D] T. Dodd (1995), *Computing: The Technology of Information*, Oxford University Press, 160 pp.

[E] G.L. Tring, ed. (1994), *Encyclopedia of Applied Physics*, VCH Publishers.

[EC] *Directive of the European Parliament and of the Council of the European Union on the Legal Protection of Databases* (1996), 96/9/E.C., March 11.

[IBM] G. McDaniel, ed. (1994), *IBM Dictionary of Computing*, McGraw-Hill, 758 pp.

[L&S] D. Longley and M. Shain, eds. (1986), *Dictionary of Information Technology*, 2nd ed., Oxford University Press, 382 pp.

APPENDIX
C

Examples of Successful International Data Exchange Activities in the Natural Sciences

The following examples of international exchange and management of data in the natural sciences cover a diverse set of activities involving international collaboration in several subdisciplines. All of them meet the minimum criteria for successful international exchange of scientific data generated from publicly funded research.

LABORATORY PHYSICAL SCIENCES DATA

Nuclear Structure Data

The Evaluated Nuclear Structure Data File (ENSDF), a mature database that has existed in electronic form for about 25 years, consists of evaluations of nuclear structure and decay data. Obtained by a variety of experiments and often spanning decades of measurements, most of the data come from primary sources, such as articles in refereed journals. The evaluations are carried out by an international network of individuals and coordinated by the National Nuclear Data Center (NNDC) at Brookhaven National Laboratory, with the international activities coordinated under the auspices of the International Atomic Energy Agency (IAEA). In addition to the United States, evaluators come from Russia, Japan, the People's Republic of China, Taiwan, Kuwait, the Netherlands, France, Sweden, Canada, and Belgium.

The evaluations themselves are reviewed before being disseminated. In the past the evaluations were submitted via magnetic tape to NNDC; more recently, almost all have been submitted via the Internet. All foreign evaluators now routinely transmit evaluations electronically and use electronic mail to communicate with the coordinators.

Although most of the ENSDF file has an 80-character column format inherited from IBM cards, a concentrated effort has been made in the past 2 years to transform the file into a more modern, relational database. Complete conversion will not be possible until a commercial database program is available that can meet the needs of such a large, diverse file, probably after another 2 to 3 years of development. So far, however, the conversion has proceeded smoothly, and it has facilitated the development of software overlay programs that give the user transparent access to the data. One subset of ENSDF, the nuclear database NUDAT, is a true relational database that can be accessed on-line.

Traditionally, the evaluations of nuclear structure data were disseminated via a monthly hard-copy journal. Now on-line access, introduced about a decade ago, is available by logging into the NNDC (guest accounts are available for individuals who want to explore the files) or via the World Wide Web.[1] On-line access is menu-driven but has limited graphic capabilities. Users can generate high-quality PostScript files of tables and level spectra, which can be downloaded to a local machine. More than 2,000 registered users from 49 countries on six continents currently have on-line access to ENSDF.

To minimize the problems with intercontinental electronic links, ENSDF has mirror sites in Vienna and Paris and plans to provide the same at the center in Obninsk, Russia, when Russia's Internet link is established. These mirror sites are maintained by the host foreign government, with support from the IAEA. While the mirror sites have minimized problems with trans-Atlantic electronic links, delays do occur with trans-Pacific links (except those with Japan) and links to South America. To complement on-line access, various modes of dissemination are being used, including CD-ROM and floppy diskettes. Although users in less developed countries may prefer hard copies, for many the cost of the printed journal is prohibitive. Therefore, to maximize international data flow, increased access to the Internet is critical.

While nuclear scientists are the most common users of ENSDF, some of the data have important applications in other areas of science. For example, radioactive decay data are used in medical physics. To facilitate access by the medical community, a dynamically generated form of ENSDF, called MIRD, was created that can be accessed on-line. As a portion of the data in the Evaluated Nuclear Data File/B (ENDF/B), ENSDF data are also used in the design of nuclear reactors and devices. Before the early 1980s the distribution of ENDF/B was restricted, because the information was considered sensitive with respect to national security. This file, which is available electronically and on tape, contains results of nuclear reaction model calculations that could be useful to other scientists, for example, stellar astrophysicists seeking to understand nuclear processes in stars.

High-energy Physics Data

Collider Detector at Fermilab Collaboration

The Collider Detector at Fermilab (CDF) collaboration presents a unique example of international data exchange and the barriers associated with the transnational flow of data. The CDF collaboration includes more than 400 scientists from 36 national laboratories and universities in the United States, Canada, Italy, Japan, and Taiwan.[2] The CDF detector itself cost hundreds of millions of dollars to construct, and the operating expenses at Fermilab are many tens of millions of dollars per year. Because of the high costs in manpower and operations, the results from this collaboration will not be duplicated.

Data at CDF are sorted into various data streams based on the physics: electroweak physics, top quark events, b quark events, events that test quantum chromodynamics, exotica, and so on. As of the end of 1995 the grand ensemble of data from CDF was about 10^8 events at 200 kilobytes per event, or 20 terabytes. This is not an easily manageable data set. The current storage medium is 8-mm tape. Working data sets are 1/1,000 to 1/100 of the total ensemble. It is estimated that at least 90 percent of the analysis is done on the Fermilab computers with data sets at Fermilab. Relatively few data are transferred electronically. The aggregate load on the network into and out of Fermilab is only about 4 gigabytes per month. In the event that data files need to be transferred to another institution, sufficient "bandwidth" is obtained with a briefcase full of 8-mm tapes.

While the data analysis is done predominantly at Fermilab, the various subgroups of the 400-member collaboration, which are organized by branch of physics and technical activities, must meet frequently. In the past, meetings were held in a central location, such as at Fermilab. More recently, videoconferencing has been used to link institutions in the United States, Japan, and Italy to facilitate discussion and analysis of the data. However, the expense of videoconferencing done by an integrated services digital network call across international borders is a major impediment to smaller institutions, particularly universities. Videoconferencing done by Internet also creates problems, since the international connections generally do not have high bandwidth. Although none of the available technology is very good at this time, videoconferencing has the potential to be the preferred method of real-time exchange and interpretation of data, so that interactive discussions of the analysis and interpretation can proceed more efficiently and cost-effectively.

With 400 individuals in a multi-institutional, multinational collaboration, the discussion and dissemination of manuscripts and technical reports could have posed a problem. However, CDF manuscripts and reports are prepared in LaTeX with PostScript output, including encapsulated PostScript figures embedded in the file. A CDF-Notes database enables note numbers to be assigned and topic/distribution categories to be selected. There is also a CDFNEWS procedure for putting the PostScript file and a brief ASCII description into a centrally accessible directory and sending notification to all collaborating institutions. Within minutes of learning about a new posting, scientists can retrieve it and have a hard copy at work or home at their convenience.

Electronic Preprints of Topical Information in Theoretical High-Energy Physics

For researchers in high-energy particle theory—including phenomenology (theoretical calculations that can be directly related to experiment), more formal quantum field theory and string theory, and lattice and computational approaches—rapid access to information has been a higher priority than the, at times minimal, filtering provided by the conventional refereeing process. Consequently, since at least the early 1970s a hard-copy preprint distribution system has supplanted conventional published journals as conveyers of topical information. With the advent of standardized word processors in the mid 1980s, together with widespread networking connectivity by the late 1980s, researchers regarded electronic transmission of prepublication information as a natural next step. When the "e-print archives" based at Los Alamos National Laboratory came on-line in the early 1990s, they were quickly adopted within these communities as the primary mode of communicating topical research information, as well as accessing longer-term archival material during the periods covered. The e-print archives have essentially supplanted established print journals in these fields.[3]

Table C.1 summarizes the distribution statistics for three e-print archives in theoretical high-energy physics (HEP) at Los Alamos. Such success has not been seen for two other experimental physics archives, especially the experimental

TABLE C.1 1995 Statistics for Three e-print Archives in Theoretical High-Energy Physics

Archive	Start Date	Number of Subscribers (numbers approximate)	Submissions per Month	Retrievals per Month	Highly Requested[a] (percent)
hep-th (abstract theory)	8/91	4,000	196	110	8
hep-lat (computational approaches)	2/92	1,000	37	30	8
hep-ph (phenomenology)	3/92	3,000	250	75	8

NOTE: The numbers are estimates and averages for 1995.
[a]The highly requested articles are those that had more than twice the average number of retrievals.

nuclear physics archive (nucl-ex), and only now is there substantial activity on the electronic archive in theoretical condensed-matter physics, which is much more closely connected to laboratory science than is high-energy theory. The users of and contributors to the HEP e-print archives include about 6,000 high-energy physicists (experimentalists and theorists), most of whom belong to the American Physical Society.

The success of the HEP e-print archives in high-energy theory could be due to several advantages shared in the subdiscipline:

- A preexisting formal hard-copy preprint distribution system, so that creating and using the electronic version did not represent a major change;
- A standardized word-processing program (LaTeX or TeX), making the electronic format easily portable;
- A subject esoteric enough that the contributions were of uniformly high quality, making refereeing less crucial to maintaining quality control; and
- A critical mass of participants accustomed to e-mail use.

Materials Science Data

Data in materials science describe a wide variety of properties (e.g., mechanical, electrical, thermal, and structural) of all types of materials (e.g., metals, insulators, and semiconductors) under many different conditions. The data also relate to differences in methods for preparing and refining materials, and, in processes such as the doping of semiconductors, to difference in methods for achieving desired levels of controlled impurity. They have little or no timeliness, except insofar as up-to-date data may replace older, less confirmed data. Hence the databases have moderate permanence, but grow largely as new materials are added and conditions and properties are extended.[4]

Although most of the collaborations involve the developed countries, international involvement in data activities is increasingly significant. Examples in materials science include the following:

- *Structure Reports*, a printed crystallographic compendium that is edited at a Canadian university, published in the Netherlands, and has contributors from all over the world;
- Science Group Thermodata Europe, a consortium of eight different laboratories from four European countries that compiles and analyzes thermodynamic data;
- The Alloy Phase Diagram International Commission, whose members from 12 countries collaborate in the compilation and evaluation of phase diagrams of metallic systems;
- The collection of diffusion data from the world literature that are abstracted, recompiled, and published in Switzerland and;

- Activities of the CODATA task group on materials database management, with members from seven countries.[5]

The potential benefits of computer access to technical information are the same for materials scientists and engineers as for researchers in other fields of technology. With reference to alloy design, three additional capabilities are seen to be of increasing importance:

- The facilitation of empirical searches for correlations of fundamental parameters from very large volumes of data through the making of two-dimensional and higher cross-plots;
- The ability to make and display arbitrary sections of three-dimensional objects (e.g., crystal structures or ternary phase diagram modes); and
- The ability to simulate crystal structures, atomic arrays, kinetic processes, and so forth, given basic information on dimensions, energetics, and modeling schema.

The most comprehensive system of on-line databases relevant to materials research is that provided by the Science and Technology Network (STN) International, which is sponsored by the Chemical Abstracts Service and by information services in Germany and Japan. Scientists and engineers are able to search approximately 20 databases covering the physical and mechanical properties of thousands of materials, as well as more than 100 factual and bibliographic databases. Although STN International provides the world's greatest concentration of data on the properties of materials, it does not include, for example, data on composites, most ceramics, semiconductors, or elastomers. The software used by STN is particularly adept at handling numeric data inquiries and is quite sophisticated in readily accommodating range searching, conversion of units, and handling of many data variables.[6] STN International is accessed via commercial on-line communication services. Searchers are charged only a modest fee for the time connected, with most of the charge based on the type and amount of data utilized in terms of the number of records involved. However, even a low fee can be prohibitive to a scientist in the developing world.

The principal advantages of a system like STN International are (1) the large number of databases available on a single system; (2) the ability to use the same software to search each of the databases; (3) the ability to search all of the databases at the same time or in groups; (4) the sophisticated search software; and (5) the ability to access only needed data without having a massive amount of data on one's own system.

Chemical Sciences Data

The work of chemists depends on many sorts of data, including compendia of text, patents, numerical tabulations, spectra rendered as analog data, and molecular structures presented both as images and in tables of bond lengths and angles. A typical textual database is the Registry of Toxic Effects of Chemical Substances.[7] Another is the bibliographic database of the Chemical Abstracts Service.[8] About 40 percent of the world's patents are for chemical substances; this information is included in DERWENT, the compendium of patents carried by STN International and other on-line services.[9]

The numerical databases incorporate thermodynamic and thermophysical data, such as heats of formation and melting points; mechanical properties, such as compressibility; transport properties, such as heat conductivity and viscosity; and kinetic data, such as rate coefficient and activation energy. Some databases, particularly the older ones, such as the Beilstein Institute's compendium of all known organic substances and its counterpart for inorganic and organometallic compounds from the Gmelin Institute, focus on the substances themselves.[10] Others, such as the database on atomic energy levels from the National Institute of Standards and Technology, catalog generic properties of a limited class of species, in this case all the atomic species for which data are available. The Cambridge structural database, originally created at the University of Cambridge and now maintained by the Cambridge Crystallographic Data Centre, an independent organization created for the purpose, is a comprehensive file of evaluated data of crystalline materials and supporting references of much interest to materials scientists and chemists.[11] Still others, such as DETHERM from the Fachinformationszentrum Chemie and SPECINFO from Chemical Concepts GmbH, carry data on specific properties of as many substances as possible—thermophysical properties for DETHERM and nuclear magnetic resonance and infrared spectra for SPECINFO. One journal, the *Journal of Physical and Chemical Reference Data*, publishes tabulations of evaluated data in conventional periodical form; these data generally reappear in large electronic databases.

Many of the large chemical databases originated in Europe, particularly in Germany, where the chemical industry and academic science thrived a century ago. The older databases appeared as bound volumes. Then loose-leaf forms became the mode of presentation. Now, most of the databases are available on-line and in computer-readable form such as CD-ROM as well.

Chemists and materials scientists who use small amounts of data to identify substances during a series of experiments need the data quickly, and so they use handbooks, local reference libraries or, when they can, on-line material. Often they need information on the quality of the data, such as the spectral resolution of tabulated infrared absorptions. They also need to be able to search databases for substances with related structures or properties. Consequently, software tools for linking data describing related substances are important for the working chemist.

The provision of many kinds of data in chemistry, materials science, and condensed-matter physics involves deriving and providing analytic representations of physical and chemical properties. This practice is important because scientists and engineers often need accurate values for properties over a range of conditions such as temperature and pressure. A significant effort in data analysis thus is directed toward finding underlying relationships on which physically and chemically sound mathematical representations can be constructed. One example is the equation of state, which links equilibrium properties such as density and internal energy with temperature and pressure.

GENOMIC SEQUENCE AND RELATED DATA

Information on DNA sequences, complete genomes of organisms, and macromolecular gene products is readily available electronically. These data have given rise to new concepts in the life sciences and to the development of new research fields, as well as to discoveries of commercial consequence such as the development of novel drugs and vaccines, diagnostic tools in medicine, improved plant varieties with better growth characteristics and improved food properties, and bacteria needed for environmental remediation. A crucial component of this capacity for innovation is large-scale international collaboration in generating, assembling, and disseminating the necessary data. The need to collect, analyze, and manage the data is leading to interdisciplinary research involving computer science and engineering, database design and artificial intelligence, and basic biological science.

Exchange of this information is accomplished through a number of databases at several institutions supported by national governmental research funding agencies, primarily the National Center for Biotechnology Information and the Genome Database in the United States, the DNA Database of Japan, the European Molecular Biology Organization, and the European Bioinformatics Institute.[12] Each organization has agreed to collect data deposited by academic, government, or industrial laboratories, put them into a transparent standard format, exchange them daily in order to maintain a common international database, and make them accessible (through the World Wide Web, by e-mail, and so on) for retrieval and analysis at any time from anywhere in the world. Databases of sequencing and protein structure information are linked through the retrieval system to other related databases (e.g., structural and molecular data, genetic maps, information on genetic diseases, and life sciences literature), so that a scientist searching for a particular DNA gene sequence can also examine the crystal structure of the protein in question and perhaps retrieve information about a related disease in humans as well.

The coordination of this database effort, the definition of standards, and planning for the future are being done by database staff and advisory committees. Currently, the international database collection contains about 500 million nucleic acid bases (the individual chemical compounds that link to make up nucleic acid (DNA or RNA) sequences) from more than 16,000 organisms. Data are being generated so rapidly that the database doubles in size every 12 months. These data are essential to ground-breaking work in molecular biology and to progress in the Human Genome Project, which is determining the sequence of the approximately 10^9 nucleotides (the combining form of the nucleic acid bases) contained in human chromosomes.[13] It is expected that by the time this effort is complete in less than 5 years, it will have a great impact on the development of diagnostic and therapeutic tools for many human diseases that are currently not treatable or whose symptoms only are amenable to mitigation.

Currently, there are no intellectual, political, or proprietary barriers limiting international access to and use of these data. The barriers are technical and economic. The most important technical barrier involves equipment and infrastructure limitations on potential end users' capability to access and then make use of the wealth of information available. These data and their free availability to researchers in the life sciences are contributing to the rapid development of new concepts and applications, and there is a great desire and consequent pressure by academic and industrial institutions to keep the data freely accessible internationally in the future.

HUBBLE SPACE TELESCOPE ARCHIVE

The Hubble Space Telescope (HST) is an example of a science program with significant international participation and open access to the data. HST was developed by NASA with the participation (nominally 15 percent) of the European Space Agency (ESA) under a memorandum of understanding negotiated between NASA and ESA. ESA also participates in its operation. HST is available for use by the international astronomy community. All science data are archived, kept proprietary (to the astronomer who proposed the observation) for 1 year, and then made available to other astronomers. The archive is accessible to the public via the Internet.[14]

Science operations for HST are centered at the Space Telescope Science Institute (STScI) in Baltimore, operated by the Association of Universities for Research in Astronomy (AURA), a university consortium, under contract to NASA. AURA has international affiliates and incorporates ESA representatives in its oversight of the STScI. ESA also contributes staff to the STScI; they are integrated into the total operation. ESA astronomers participate in HST committees and advisory structure. In addition, ESA operates a small Space Telescope (HST) European Coordinating Facility (the ST-ECF) in Garching, Germany, in collaboration with the European Southern Observatory.

HST observing is open to all astronomers worldwide via a peer review system. Under the memorandum of understanding, astronomers from ESA member countries are entitled to 15 percent of the observing time on average. In practice, they receive at least this amount through the normal peer review system.

All HST data are received by the STScI. They undergo routine processing and calibration, and both the calibrated and uncalibrated data and the engineering and other ancillary data are archived. The primary archive for HST data at the

STScI contains about 2 terabytes of data and is growing at the rate of a gigabyte per day. A duplicate copy of the science data archive is transferred to the ST-ECF, and a third copy of nonproprietary data is maintained by the Canadian Astronomy Data Center in Victoria, British Columbia. Each of these data centers also archives different sets of related data. Under NASA policy for HST, nonproprietary data are freely available, but requests for large amounts of data must be approved by NASA headquarters and are subject to a negotiated level of cost recovery, typically the marginal cost of reproduction.

Data analysis software appropriate for HST users was developed and is maintained by the STScI and is freely distributed to astronomers. It operates in a portable data analysis environment, the Image Reduction and Analysis Facility (IRAF), developed by the National Optical Astronomy Observatories. Although other large (and small) data analysis systems exist, IRAF has been adopted by a number of astronomy projects and is used by a large portion of the astronomy community both in the United States and in other countries. It incorporates both general astronomy-oriented data analysis tools and specific packages for individual observatories and facilities.

GEOPHYSICAL DATA

The World Data Centers

In the Earth sciences, with the impetus of the International Geophysical Year in 1957, the World Data Centers (WDCs) were set up under the aegis of the International Council of Scientific Unions (ICSU).[15] Their function was to provide international access to various types of observational geophysical data. This effort was very successful, and geoscientists since that time have taken advantage of the WDCs as an effective mechanism for the exchange of data. The WDCs circumvented what otherwise would have been insurmountable political barriers to exchange of scientific data in the era of the Cold War and allowed scientists from the East and the West to use data collected by both sides. The protocol for the WDCs was that any scientist could obtain any of the data residing in the WDCs without government or other restrictions. Of course, only subsets of data collected in different countries were placed in the WDCs, but substantial amounts were made available. Initially, the data were in analog form, but in recent years data holdings have been archived and disseminated in digital form (e.g., via tapes), and additional WDCs have been established for different types of geophysical data. Increasingly, users can browse the data holdings and receive data via electronic networks.

In the United States, national data centers, such as the National Geophysical Data Center operated by the National Oceanic and Atmospheric Administration (NOAA), serve a dual role, with a subset of their holdings designated as a WDC and therefore available to any user, domestic or foreign. Other examples include the U.S. Geological Survey's Earth Resources Observation Systems (EROS) Data Center, which houses the newly established WDC for land remote sensing, where a subset of remote sensing data is made available,[16] and the Department of Energy's Carbon Dioxide Information Analysis Center, which also serves as a WDC for trace gases in the atmosphere.[17] Although the WDCs provide one very effective avenue for the transnational flow of geophysical data, many important observational data sets are not available through the WDCs and must be obtained through other means, some of which are discussed below.

Seismic Data

Many thousands of seismic events occur throughout the world each year, some large and destructive. The detection and location of earthquakes and determination of their magnitudes require a globally distributed network of well-calibrated, sensitive seismic stations that continuously record ground motions. Such a network necessarily involves seismic stations in many countries around the world. Data are gathered by a combination of individual institutions and different regional and global networks operated by individual organizations under agreements with countries or institutions where the stations are located. The determination of the location, depth, time of occurrence, and magnitude of an earthquake makes use of data from ground motions observed by many stations, at different distances and azimuths from the source. Monitoring of global seismic activity therefore involves the transnational flow of data and information both in real time and on a recurring basis.

Global seismic monitoring serves purposes other than earthquake hazard assessment and mitigation, the most important being enforcement of international treaties governing underground nuclear explosions. Underground explosions are recorded by the same seismic stations that record earthquakes; like earthquakes, they can be detected and located. Considerable past and current research is devoted to developing reliable methods for distinguishing underground nuclear explosions from natural earthquakes and from mining blasts. Under protocols currently being developed among participating countries, all parties are to have equal access to continuous real-time recordings from approximately 50 seismic arrays and from many additional single stations distributed around the world. Each country will then carry out its own assessment of recorded events. The transnational flow of these data in near-real time will be formalized as part of the Comprehensive Test Ban Treaty.

In the early part of this century, the international exchange of seismic data was accomplished by a scientist writing to the seismologist operating each station of interest and asking to borrow the original (analog) recording of the event being studied. When the work was complete, the users returned the original recordings to the respective station operators. By contrast, a recently implemented data access capability at the Incorporated Research Institutions for Seismology (IRIS)

Data Management Center (DMC) allows any scientist (U.S. or international) to download via the Internet the signals recorded at approximately 20 global seismic stations within about 1 hour of an event's occurrence (magnitude greater than 5).[18] Similarly convenient Internet access to continuous recordings of many other international stations is possible through the DMC, but with a time lag necessitated by not having real-time or near-real-time data transmission from some of the available stations.

IRIS's DMC has become the international Federation of Digital Seismic Networks' archive for continuous digital data. Global digital seismic data from stations distributed around the globe are available through the DMC. Users can browse electronically to determine what data are available and can place requests for data sets they wish to receive; their requests are filled and the data transferred either electronically via the Internet (for modest-size data sets) or via high-density media such as Exabyte cassettes (for large requests). The DMC also serves as a broker for individuals who wish to obtain data from foreign stations that are not routinely archived at the DMC. This valuable service is accomplished by means of data transfer links to data archives in other countries; users would otherwise have to access and transfer data from these various sources individually. In this way, the DMC operates as a "virtual" data center from which the user extracts desired data, some of which do not physically reside at the center.

The World Weather Watch

The World Weather Watch is the most formally organized international global observation, communication, processing, and archiving system at this time.[19] This distinction stems from the early recognition that scientific understanding and prediction of the atmosphere, even for only a day or two in advance, require observations from very large areas. Beginning more than 100 years ago, the observations were sent by communication systems in near-real time through internationally agreed upon arrangements and procedures. For the last several decades, data have been processed and archived on a global basis through a system of world and regional meteorological centers and world and regional data centers for meteorology and oceanography. During this period the World Weather Watch has developed many of the characteristics required for an effective system for international exchanges of scientific data and therefore can be considered one of the primary models for other such systems.

The development of the World Weather Watch was accelerated in the 1960s as a result of the potential capability of Earth-orbiting satellites to obtain atmospheric and oceanographic data on a global basis, and the advent of computers capable of handling large volumes of diverse data for numerical weather predictions on a global basis. An extensive planning and coordination process was put in place in the World Meteorological Organization (WMO) to expand the global observing component of the World Weather Watch through polar and geostationary satellites and additional in situ observation, to develop an improved telecommunications system capable of exchanging data in real time among all nations, and to establish a system of supporting data centers. Three such centers, in Washington, D.C., Moscow, and Melbourne, were established in the mid-1960s, along with regional meteorological centers to serve specific continental and oceanic areas. These World Meteorological Centers are responsible for the preparation and distribution of an agreed upon set of global products to all nations, through the Global Telecommunications System of the World Weather Watch. Similarly, the regional meteorological centers prepare products as agreed for their specific areas of responsibility. The archival, storage, and retrieval systems for retrospective use of the data are maintained by the World Data Centers, and, by virtue of recent expansion, the Regional Data Centers.

WMO does not operate any observing stations, telecommunication systems, or processing centers; through its member nations WMO is responsible only for the planning and coordination of the World Weather Watch. This includes developing the scope and extent of the observing networks, the characteristics and standards of the telecommunication systems, and the products to be prepared at the centers. The World Weather Watch, therefore, is built on the national meteorological systems of each member nation. The national meteorological system in the United States is quite extensive because of the great impact of weather, especially severe storms such as hurricanes, tornadoes, blizzards, and flash floods, on people and industry. Services are provided through a public-private partnership. The federal government is responsible for public forecasts and forecasts related to the safety of life and protection of property—severe weather and flood warnings for the country and surrounding oceans, and forecasts and advisories for aviation terminals and en route paths. The private sector provides tailored forecasts for specific clients, and through television, radio stations, and, to an increasing degree, the Internet leads in the dissemination of severe weather and flood warnings to the public.

The federal government operates an extensive satellite and ground-based observing system, together with meteorological prediction and data centers, to obtain the data and products needed to carry out its responsibility for providing services. These data and products are made available to the private sector with no restrictions and at low incremental costs. These same data and products are used in fulfilling the requirements and agreements within the World Weather Watch and for research internationally and nationally. For example, the data from U.S. geostationary satellites can be received directly by centers in South America; the data from U.S. polar-orbiting meteorological satellites are distributed on the World Weather Watch Telecommunication System; and the products prepared by the National Meteorological Center near Washington, D.C., which are designed to meet national requirements, are provided to all countries through the World Weather Watch. Likewise, the data are available for all research programs, primarily through the NOAA National Climatic Data Center and the National

Oceanographic Data Center. Other nations have functioned in a similar way—the same centers fulfilling both national needs and international commitments.

The close interaction between operational meteorological services and research in the atmospheric sciences, nationally and internationally, has proved extremely effective. During the late 1960s and 1970s, an extensive program—the Global Atmospheric Research Program—was undertaken internationally by ICSU and WMO to improve the accuracy and extend the time range of weather forecasts. A joint mechanism was established within which the ICSU scientists led the planning of major observational field experiments and WMO led the implementation through national contributions of member nations. The largest and most complex was a global observational experiment during which the World Weather Watch Global Observation System was augmented with additional observations from ships, aircraft, and satellites to provide the most comprehensive set of global observations ever acquired. Again, the World Data Centers were responsible for archiving the data from the experiment for use in the associated research programs. Simultaneously during this period, the data from the World Weather Watch were being used by the U.S. government and private sector to provide services. Such multipurpose use of meteorological data has historically been very effective and efficient.

However, the traditionally unrestricted exchange of data in meteorology has been placed in jeopardy in recent years. Pressure on weather services from some governments to charge users for services other than public weather forecasts and severe weather and flood warnings has led to proposals to place restrictions on the use of data and to charge substantially for real-time data or data sets between data centers. The meteorological services in Western Europe have been the most aggressive in charging industries and organizations for specialized services and private meteorological companies for data. This situation was a major consideration at the meeting of the WMO Congress in 1995, which adopted an understanding by members of the WMO that they would endorse the free and unrestricted exchange of data for research and education, and for an agreed set of data—satellite and in situ—to be exchanged in real time. However, it included a provision that an individual country could place restraints on data made available beyond the agreed level,[20] and this has resulted both in a reduction of data freely available for research as well as in significant adminstrative expenses.

NOTES

1. See <http://www.nndc.bnl.gov >.
2. See <http://www-cdf.fnal.gov/> for additional information on the Collider Detector at Fermilab.
3. See <http://xxx.lanl.gov/> for the e-print archives.
4. Summaries of data sources for materials science and engineering (both print and electronic) have been published. See H. Wawrousek, J.H. Westbrook, and W. Grattidge (1989), "Data Sources of Mechanical and Physical Properties of Engineering Materials," *Physik Daten*, 30-1, Fachinformationszentrum, Karlsruhe, Germany; J.H. Westbrook and W. Grattidge, eds. (1988),

"The CODATA Referral Database (CRD)," based on the CODATA Database Directories and the 1988 revision of the UNESCO "Inventory of Data Referral Sources in Science and Technology," available from CODATA, 51 Boul. De Montmorency, Paris; J.H. Westbrook (1986), "Materials Information Sources," *Encyclopedia of Materials Science and Engineering*, M.B. Bever, ed., p. 527, Pergamon; F.C. Allan and W.R. Ferrell (1989), *Database 12*,(3):50-58; M.K. Booker (1986)"Computerized Materials Databases," *Encyclopedia of Materials Science and Engineering*, pp. 796-800, Pergamon. The role of the computer in accessing and manipulating materials data for alloy design is discussed by Westbrook in J.H. Westbrook (1993), "Data Compilation, Analysis, and Access: The Role of the Computer," *MRS Bull.*, 18:44-49. R.A. Matula (1989), "The Importance of Numeric Databases to Materials Science," *J. Res. Natl. Inst. Stand. Technol.*, 94:9-14, emphasizes the importance, in an industrial setting, of computer access to numeric databases in materials science.

5. Functioning almost entirely without external financial support and on a volunteer basis, this CODATA task group coordinates work in this field, promotes standards, communication, and awareness; assists in education and training; and publishes an international register of materials database managers.
6. See <http://www.cas.org/stn.html>.
7. See <http://www.rs.ch/krinfo/products/datastar/sheets/RTEC.htm>.
8. See <http://www.cas.org> for information about the Chemical Abstracts Service.
9. See <http://www.derwent.co.uk>.
10. Access fees to these are often prohibitive. Several respondents to the committee's "Inquiry to Interested Parties" (see Appendix D) noted specifically that they would like access to the Beilstein databases but considered them too costly.
11. See the Cambridge Crystallographic Data Centre home page at < http://csdvx2.ccdc.cam.ac.uk/ccdchome.html>.
12. See the National Center for Biotechnology Information home page at <http://www.ncbi.nlm.nih.gov/>.
13. See <http:/www.nghgr.nih.bov/HGP/> for additional information regarding the Human Genome Project.
14. See <http://www.stsci.edu/archive.html> for additional information about the Hubble Space Telescope archive.
15. See <http://www.ngdc.noaa.gov/wdcmain.html> for a description of the World Data Centers System.
16. See <http://edc.www.cr.usgs.gov> for the EROS Data Center home page.
17. See <http://cdiac.esd.ornl.gov> for the Carbon Dioxide Information Analysis Center home page.
18. See <http://www.iris.washington.edu/dmc.new.html> for the IRIS Data Management Center home page.
19. For additional information on the World Weather Watch, see the World Meteorological Organization home page at <http://www.wmo.ch:80/www/www.html>.
20. See R.S. Greenfield, E.W. Friday, and M.C. Yerg (1995), "WMO Adopts a Resolution Governing the International Exchange of Meteorological and Related Data and Products," *Bulletin of the American Meteorological Society*, 76(8):1478-1479.

APPENDIX
D

Inquiry to Interested Parties on Issues in the Transborder Flow of Scientific Data

Dear Colleague:

The U.S. National Academy of Sciences/National Research Council (NAS/NRC) is undertaking a study to review important issues and trends in the international flow of scientific data, particularly along transborder electronic networks. The study will characterize the technical, legal, economic, and policy issues that have an influence (favorable or negative) on access by the scientific community to scientific data. The scope of the study includes symbolic and substantive textual data as well as numerical data; bibliographic data are only included to the extent that they are related to substantive and numerical data. The study will identify and describe both the positive aspects and the barriers or hindrances that have impacts on research in the natural sciences (physical, astronomical, biological, and geological) and across those disciplines. These will be illustrated by representative examples. Finally, it will identify medium- and long-term trends likely to have significant discipline-specific and interdisciplinary influence on the access to and use of scientific data, particularly in electronic forms, and, where appropriate, suggest approaches that could help overcome barriers and hindrances in the international context.

The attached "Inquiry to Interested Parties" is a tool to help us identify significant issues and provide important information to us from the viewpoints of data users and suppliers regarding transborder dissemination of and access to scientific data in the natural sciences from the legal, policy, economic, and technical perspectives. Because of the nature of this inquiry and the means by which it is being distributed (i.e., not a demographically controlled sample), we do not intend it to be a survey base for a statistical study. Rather, we are interested in facts, interpretations, opinions, and real examples that will help us gain insight into the main issues of the study. We also are seeking illustrative material that we can use to communicate the situation to the scientific and governmental establishments.

The goal of our study is to help improve access to scientific data and services internationally. We therefore hope that your interests are common with ours, and that you will assist us by providing your views on these issues by taking some time to fill out and return this form. We recognize that not every respondent is likely to be able to comment on every question, and we do not wish to have the specific questions to be a limit on what you wish to inform us about. Therefore, please skip any questions that you do not feel you can address meaningfully, and add any points that you would like for us to know or consider. Feel free to use additional pages or attach other pertinent information if you have more that you wish to say to us.

Please send your response and any related documentation by 31 January 1996 to:

Paul F. Uhlir
Director, U.S. National Committee for CODATA
National Academy of Sciences/National Research Council
2101 Constitution Avenue, N.W.
Washington, D.C. 20418 U.S.A.
Telephone: (202) 334-3061; Fax: (202) 334-2154
Internet: BITS@NAS.EDU

We very much look forward to hearing from you.

Sincerely,

R. Stephen Berry
Study Chairman

INQUIRY TO INTERESTED PARTIES ON ISSUES IN THE TRANSBORDER FLOW OF SCIENTIFIC DATA

Please provide the following information:

Name:

Address:

Telephone/fax/e-mail *(optional)*:

Brief description of your data activities and discipline background:

Are you answering this questionnaire as a scientific data: user ( ), producer ( ), distributor ( ), vendor ( ), system manager ( ), network operator ( ), policymaker ( ), or other _____ ? [Please check all that apply.]

1. **Barriers to Data Access.** Some restrictions on access to scientific data frequently are considered necessary to protect various interests as well as the integrity of the data. In your experience, have restrictions on data been a problem? Can you identify any specific impacts or trends? Please explain.

2. **Pricing of Data.** If you *use* data for scientific research, please tell us:
   (a) What data sets you have recently used for which you or your institution paid nothing, and in what form did you get these data (e.g., WorldWideWeb, other on-line, CD-ROM, diskette, tape, film, paper, etc.)?

   (b) What data have you recently used for which you paid any amount (including the cost of reproduction or communication connectivity); in what form did you get these data, how were you charged (e.g., flat rate, charge per use, etc.), and how much?

   (c) What data would you like to use for your research, but consider them too expensive/costly? What is the cost of such data and what is their value (apart from cost)?

   (d) For the data listed under (c) above, what arrangements could help make these data available to you? In what form would you like to be able to get these data?

If you *supply* data for scientific research (and perhaps for other uses), please tell us:

(e) Are you a profit-making enterprise; if not, what is the form and intent of your organization?

(f) What kind of data do you supply that are used by scientific researchers?

(g) Besides scientific researchers, what kind of other users of your data are there, if any?

(h) Do you provide special pricing for research/academic users? If so, what is your pricing policy?

(i) What are the media you use to distribute your data (e.g., paper, film, tapes, diskettes, CD-ROMs, on-line, etc.)?

(j) If you sell or otherwise market your data, what is your perception of the price elasticity and demand for the data you distribute? What changes would you make to your data products and services if demand were to increase?

3. **Protection of Intellectual Property.**

   (a) What are the principal legal and technical mechanisms actually used for protecting unauthorized uses of data in your country/institution/discipline area?

   (b) Can you provide any information about how such legal or technical mechanisms are implemented or enforced? What are the positive and negative impacts?

4. **Less Developed Countries.**

   (a) In your experience, what have been the principal problems associated with transferring data into or out of "less developed countries," including those nations from the former Soviet Union?

   (b) What can be done to help alleviate these problems, especially by the international scientific community?

5. **Electronic Networks.**

   (a) Has the development and growth of the Internet and other electronic networking services affected the way you access or distribute data internationally? Please give specific examples if you can.

   (b) How do you think the situation with electronic networks will change in the next 5-10 years or so, and what are the likely impacts to your activities?

6. **Other Technical Issues.**
   (a) Besides those associated with electronic networks, what are the most important technical benefits or problems you have experienced in either disseminating or accessing data internationally?

   (b) What changes do you anticipate over the next 5-10 years, and what are the likely impacts to your activities?

7. **Scientific Data for Global Problems.**
   (a) In your view, what is the role of international scientific data for addressing global problems, now and in the future? Please elaborate.

   (b) What can be done to enhance the availability or exchange of scientific data to better address these concerns?

8. **Other Issues.** Do you have any specific concerns or examples of successes that you believe should be considered in this study? In addition, we would welcome your suggestions for other institutions or individuals to contact with regard to these questions, as well as any references to key documents.

[Note: A summary of the most useful responses will be available on the U.S. National Committee for CODATA Web site at: <http://www.nas.edu/cpsma/codata.htm>.

Index

A

Access to scientific data. *See* International access to scientific data; Data, unrestricted access
Advanced Very High Resolution Radiometer (AVHRR), 58, 123
Africa, 42, 46, 70, 107
African Data Dissemination Service (ADDS), 107
Agencies. *See* Federal government science agencies
Agreement on Trade Related Aspects of Intellectual Property Rights (TRIPS), 150, 154-156, 161, 170, 178, 183-184
Alloy Phase Diagram International Commission, 209
American Association for the Advancement of Science, 92, 94, 109
American Astronomical Society, 66
American Geophysical Union, 64
American Institute of Physics, 64
American Meteorological Society, 64
American Society for Mechanical Engineers, 96
Animation, 4, 57, 65, 67-68, 196
Anticompetitive exclusive property rights. *See* Exclusive property rights
Antitrust law, 142, 164

Applied research, 1, 18, 20, 133
Archives, 5, 59-60, 72, 82-83, 196. *See also* Long-term data sets; Retention, long-term; Retrospective data sets
Argentina, 54
Association for Progressive Communications, 42, 46
Association of American Geographers, 64
Association of Universities for Research in Astronomy (AURA), 117, 213
Astronomical Data Center, 66
Astronomical sciences, 1, 19, 34, 53-54, 66
Astronomical Society of the Pacific, 66
Astronomy and Astrophysics, 66
Astrophysical Journal, 66
Astrophysics Data System (ADS), 66, 71
Asynchronous transfer mode (ATM), 25, 32, 35, 196
Atmospheric sciences, 55, 57, 78, 216-218
Australia, 66
Australian Oceanographic Data Centre, 118
Authentication issues, 13, 43
Automatic license, 164

B

Balancing divergent interests, 2, 139-145, 162, 164-170

Barriers to access, 3, 4, 8, 76, 89. *See also* Economic factors and trends; International access to scientific data; Legal infrastructure
Basic research, 1-2, 7, 9, 17-18, 97, 133, 163-164
Beilstein databases, 61, 211
Belgium, 205
Berne Convention for the Protection of Literary and Artistic Works, 9, 135, 145, 155, 160-161, 170, 183
"Big science." *See* "Megascience" programs
Bilateral agreements, 81
Biological data, 56, 72-74, 84
Biological sciences, 1, 5, 11, 19, 56-57, 73, 83-88
Books. *See* Publishers, commercial
Brazil, 54
Broadband transmission, 28, 30, 197
Bromley Principles, 79, 82
Brookhaven National Laboratory, 63, 205
"Browsing," 29, 143, 197

C

Cambridge Crystallographic Data Centre, 211, 219
Canada, 206
Carbon Dioxide Information Analysis Center (CDIAC), 71, 105, 215, 219
Center for Information and Numerical Data Analysis and Synthesis (CINDAS), 119
Centers for Disease Control and Prevention (CDC), 120
Chemical Abstracts Service, 211, 219
Chemical sciences, 50-51, 210
Chile, 41
China. *See* People's Republic of China
Circular A-130, 126, 131, 163, 187
Climate. *See* Global climate change
CODATA. *See* Committee on Data for Science and Technology
CODATA Commission on Standardized Terminology for Access to Biological Data Banks, 11, 85, 88, 101
CODATA Task Group on Fundamental Constants, 88
CODATA Task Group on Outreach, Education, and Communication, 108
Cold War, end of, 4, 21, 77, 111, 214
Collaborative research, 6, 14, 21, 29, 31-32, 45, 63, 97-100, 207-208

Collider Detector at Fermilab (CDF), 207, 218
Commercialization of data, 29, 31, 99, 111, 168-169. *See also* Markets
Commercial users, 120, 167
Commission of the European Communities (CEC), 8, 145-148, 154-155, 164
Committee on Data for Science and Technology (CODATA), viii, 1-2, 19, 85, 88, 219
Committee on Earth Observation Satellites (CEOS), 94-95
Committee on Geophysical and Environmental Data, 79
Committee on Science and Technology in Developing Countries (COSTED), 108
Communications. *See also* Wireless communication
  costs of, 27-28, 41, 42, 44
  two-way, 6, 13
Compact Disk-Read Only Memory (CD-ROM), 34, 45, 64, 98, 126, 141
Competitiveness, 7, 9, 116, 148, 162. *See also* Exclusive property rights; Unfair competition laws
Compton Gamma Ray Observatory, 60, 81
Compulsory license, 147, 159, 169
Computational Materials Science, 67
Computer science research, 13, 43
Computers, 13, 40-42, 44, 102, 110. *See also* Personal computers
  hybrid analog/digital, 26
  scientific requirements for, 36, 39
Computing, costs of, 27-28
Conflicts of law. *See* Legislation
Congressional Budget Office, 58-61, 104
Consortium for International Earth Sciences Information Network (CIESIN), 94, 99, 108
Constants. *See* Fundamental constants; Natural constants
Constitution. *See* U.S. Constitution
Consultative Group on International Agricultural Research (CGIAR), 107-108
Contracts, 135
  harsh terms of, 159
Copying. *See* Photocopying
Copyright Act of 1976, 140, 143-144
Copyright law, 16, 48, 136-140, 143-144, 156. *See also* Fair use; Noncopyrightable databases; Royalties

INDEX

Costs. *See* Computing; Data; Reproduction and distribution
Council of Ministers, 148, 152, 154-156
Cyberspace, 9, 143, 145, 161, 197

D

Data, 4-6, 21-22. *See also* Networks
  accessing (*See* International access to scientific data; Full and open exchange of data)
  acquiring, 82
  analog, 196
  collecting, 27-28, 30, 62, 198
  compatibility of, 5, 88-90
  cost of, 6 (*See also* Data, pricing)
  declassification of, 77-78
  describing, 36, 38, 49-51
  digitizing, 2, 11, 73, 101, 111
  distributing, 4, 6-7, 62-64, 114-116, 129
  evaluating, 199
  exchanging, 21, 83, 198
  growing volume of, 4, 58
  historical, 11, 71, 101
  indexing, 36, 38
  interfacing, 75
  managing, 10, 11, 13, 61-64, 198
  modeling, 65, 70
  nondiscriminatory availability, 79
  nonproprietary, 79
  preserving, 57, 61-62, 71
  pricing, 7, 14, 124-126
  primary, 21, 49, 202
  privatizing, 6-7, 14, 111, 116, 120-124, 202 (*See also* Intranets)
  processing, 50, 198
  providing (*See* Sole-source data providers)
  purging, 82
  raw, 202
  remote access, 203
  rescuing, 13, 98
  retaining (*See* Data, storing)
  retrieval, 83, 206
  sources of, 11
  spectral, 203
  storing, 4, 21, 38, 82, 110
  transferring, 198
  transnational exchange of, 19, 21, 83, 97-100
  unique, 163
  universality of, 48
  unrestricted access, 83

  using (*See* Documentation of data sets)
  validating, 198, 204
  verifying, 100, 204
Database industry, 153
Database publishers. *See* Publishers
Databases, 199. *See also* Genome databases; Noncopyrightable databases
  commercial value of, 168
  converting, 89
  electronic, 8
  "insubstantial parts of," 152, 157-158
  managing, 198
  relational, 203
  vulnerability of, 140-142
Data centers, 5, 21, 60, 83
Data integrity, 198
Data Management Center (DMC), 216
Decryption, 198. *See also* Encryption
Deoxyribonucleic acid (DNA), 48, 56, 73, 212
Derivative work rights, 139, 151
DETHERM, 211
Developing countries, 6, 12, 40-42, 90-100
  accessing data from, 97-100
Digital versus analog data, 20
Digital versatile disk (DVD), 34, 45
Digitizing. *See* Data
Directive on the Legal Protection of Databases, European, 9, 142, 148-160
Distributed networks of data centers, 83
Documentation of data sets, 5, 11, 76-77, 106

E

eApJ. *See Astrophysical Journal*
Earth Interactions, 64
Earth Observation Satellite Company (EOSAT), 121
Earth Observing System (EOS), 30, 45, 80
Earth Resources Observation Systems (EROS), 57, 58, 62, 106, 118, 215, 219
Earth sciences, 5, 54-56, 69, 75
Earth system processes, 6, 54, 97
Earthquakes, 55. *See also* Seismic arrays and stations
Ecological Society of America, 64
Economic factors and trends, 1, 6-8, 110-131, 132-133
  decreasing computing costs, 3, 27
  restrictions and barriers, 3, 91-92
Education. *See* Science education
Electronic publication, 64-65, 142, 165, 199

Electronic storage media, 4, 21
  rapid obsolescence of, 36, 38, 72, 74
Encryption, 43, 156, 159-160, 199
Engineering applications, 51
England. See United Kingdom
Environmental sciences, 5. See also
  Observational environmental sciences
E-print archive, 199, 208-209
European Bioinformatics Institute, 212
European Community, 93, 134, 145-148, 154
European Dictionaire Automatique, 33
European Directive on Databases. See Directive on the Legal Protection of Databases
European Molecular Biology Organization, 212
European Southern Observatory, 54
European Space Agency (ESA), 54, 118, 213
European Space Information System, 117
Evaluated Nuclear Data File/B (ENDF/B), 119, 206
Evaluated Nuclear Structure Data File (ENSDF), 52, 205-206
Exclusive property rights, 136-137, 146
Extraction, 157-158
  unauthorized, 158
  unfair, 147
"Extraction right," 154
  social costs, 154, 162

F

Fair use, 8-9, 12, 15-16, 143-144, 167-168, 171
Fatty Acid Methyl Ester (FAME) system, 120
Federal government science agencies, 2-3, 13, 20, 44. See also *individual agencies*
Federation of Digital Seismographic Networks (FDSN), 59, 216
Feist Publications, Inc. v. Rural Telephone Service Co., 8, 16, 139, 154
Fiber-optic communication, 24, 26
Fidonet, 42
File transfer protocol (FTP), 131
First Amendment. See U.S. Constitution
Flexible image transport system (FITS), 34
Food and Agricultural Organization (FAO), 86, 93
Foreign aid, 12-13, 44, 92
France, 206
Freedom of scientific inquiry, 2, 17
Full and open exchange of data, 3, 7, 9-10, 15, 17, 22, 81-82, 101, 199
Fundamental constants, 56
Fundamental research. See Basic research

G

Gemini project, 54
Genbank, 117
Genome databases, 56, 117, 212-213
Geographic Information System (GIS), 75-76
Geological sciences, 1, 19. See also Earth sciences; Environmental sciences; Observational environmental sciences
Geophysical data, 75, 214-218
Global Atmospheric Research Program, 218
Global Atmospheric Watch, 69
Global Change Research Information Office, 55, 69, 79, 82
Global climate change, 6, 55
Global Climate Observing System, 69, 105
Global data sharing. See International access to scientific data
Global observational data sets, 97
Global Ocean Observing System, 69, 105
Global Terrestrial Observing System, 69-70, 105
Government science agencies. See Federal government science agencies
Great Britain. See United Kingdom
"Group of Seven" nations, 10, 101

H

High density-read only memory (HD-ROM), 45
High Energy Astrophysical Observatory 2, 60
High-energy physics (HEP), 206-208
High-performance parallel interface (HIPPI), 30
Historical data. See Data
House. See U.S. House of Representatives
H.R. 3531. See U.S. Database Investment and Intellectual Property Antipiracy Act
Hubble Space Telescope (HST), 4, 60, 110, 113, 117, 120, 213-214, 219
Human Genome Project, 4, 23, 27, 120, 213, 219
HyperText Mark-up Language (HTML), 33-34

I

Icarus, 66
Iceland, 183
Image Reduction and Analysis Facility (IRAF), 214
Incompatibilities. See Standardization

Incorporated Research Institutions for
    Seismology (IRIS), 57-59, 119, 215-
    216, 219
Incremental cost pricing, 14, 125-126
Indonesia, 42
Information Infrastructure Task Force (IITF)
    White Paper, 143, 155-156, 159-161
Information Market Policy Action (IMPACT)
    program, 145
Information technology, 3-4, 24, 110
    trends in, 24-46
Infrared Astronomical Satellite, 60
Innovations, 1, 133, 136, 140
    subpatentable, 137
In situ measurements, 75, 97
Institute of Physics (U.K.), 64
"Insubstantial parts." *See* Databases
Intangible property, 161
Integrated services digital network (ISDN)
    services, 32
Intellectual property rights (IPR), 5, 8-10, 48, 132-188
Interdisciplinary research, 74-76
Intergovernmental Oceanographic Commission, 70
Intergovernmental organizations, 63, 93. *See also individual organizations*
International access to scientific data, 2, 40-42, 90-100
    developments affecting, 2-10, 17
    safeguarding, 7-10, 166-169
International Atomic Energy Agency (IAEA), 52, 205
International Committee on the Taxonomy of Viruses, 85
International cooperation, 19, 83
International Council of Scientific Unions (ICSU), 10-13, 44, 55, 88, 99, 101-103, 108, 214, 219
International Decade of Natural Disaster Reduction, 75
International Foundation for Science (IFS), 108
International Geosphere-Biosphere Programme (IGBP), 4, 22, 27, 55, 69, 77, 106
International Organization for Standardization (ISO), 46
International organizations, 6, 12, 93. *See also individual organizations*
International research, 58-61
International Research and Exchange Board (IREX), 94, 109

International Science Foundation, 109
International Telecommunications Union, 13, 44
International Ultraviolet Explorer, 60
International Union of Biological Sciences, 85
International Union of Microbiological Sciences, 88
International Union of Pharmacology, 85
Internet, 3, 29, 31, 42, 65, 68, 200. *See also* World Wide Web
    congestion on, 8, 26, 35-37, 126-128
Internet Engineering Task Force (IETF), 12, 32, 34, 43
Internet Protocol Next Generation (IPng), 35, 45
Internet II, 13, 35
Intranets, 4, 29, 35, 200
Investment, 9, 154, 168. *See also* Profit; Return on investment; "Substantial investment"
    incentives for, 141
Italy, 207

J

Japan, 205
Journals. *See* Scientific journals

K

Kenya, 42
"Knowbots," 26, 33, 200
Kuwait, 205

L

Laboratory physical sciences, 5, 51-52, 205
    data compatibility in, 88-90
    data exchanges in, 205-212
Land Remote Sensing Policy Act of 1992, 121
Landsat, 77, 111, 121-123, 163
Language translation, 33
Large research programs. *See* "Megascience" programs
LaTeX, 65, 88, 200, 207-208
Laws. *See* Legislation
Legal infrastructure, 1, 8-10, 139
    restrictions and barriers, 3, 9, 136
Legislation, 132. *See also* Antitrust law; Copyright law; Trade secret law; Trademark laws; Unfair competition laws; *individual laws*

conflicts in, 161
pending, 9-10, 160-161, 171
Legislative process, 8, 15
Liability principles, 137, 148, 164
Liability regime, 136
Libraries, 8, 145
Licenses, 151, 154. *See also* Automatic license; Compulsory license
Licensing fees, 165
Liechtenstein, 183
Long-term data sets, 82. *See also* Archives; Retention, long-term; Retrospective data sets
Los Alamos National Laboratory, 30-31, 45, 208

M

Machine translation, 29
Manufacturing applications, 18, 51
Marginal cost pricing, 7, 14, 126
Market forces, 3, 162-164, 168
Market power, 151, 164
　abuses of, 154
Market price system, 113
Markets, 9, 111-114, 139, 140, 162-164. *See also* Competitiveness
Materials sciences, 67, 106-107, 211
Media. *See* Electronic storage media; Print media
"Megascience" programs, 19, 23, 58-61
Metadata, 106, 201. *See also* Documentation of data sets
Metals Information Analysis Center (MIAC), 119
Meteorology, 79, 216
Mexico, 76, 98
Microbiology, 96
Micro-electromechanical systems (MEMS), 25
Mission to Planet Earth (MTPE), 193
Modeling and Simulation in Materials Science and Engineering, 67
Molecular biology, 67, 212
Monopolies
　profit-making, 7, 151-152, 158, 162
　public, 116
　regulated, 124
　unrestricted, 7
Moore's Law, 28, 44, 201
Multicast broadcast backbone network (MBone), 31-32, 45
Multinational authorship, 21

N

National Aeronautics and Space Administration (NASA), 34, 45, 54, 64, 80-81, 93, 117, 213-214
　data availability policy, 80-81
National Cancer Institute, 41
National Center for Atmospheric Research, 117
National Center for Biotechnology Information (NCBI), 212, 219
National Climatic Data Center (NCDC), 72, 119, 217
National Geophysical Data Center (NGDC), 72, 214
National Human Genome Research Institute, 22
National information infrastructure, 193
National Institute of Standards and Technology (NIST), 89
National Institutes of Health (NIH), 41, 117
National Land Remote Sensing Satellite Data Archive, 62
National Library of Medicine (NLM), 41
National Nuclear Data Center (NNDC), 52, 63-64, 206
National Oceanic and Atmospheric Administration (NOAA), 62, 98, 116, 119, 121, 214, 217-218
National Oceanographic Data Center (NODC), 98, 217-218
National Optical Astronomy Observatories, 214
National Research Council (NRC), 20, 38, 45-46, 54, 68-69, 74-75, 103-109
National Science Foundation (NSF), 31, 54, 85
National security issues, 77
National Space Policy 1996, 62, 79
National Space Sciences Data Center, 117
Natural constants, 50
Natural language processing, 4, 29, 32-33
Natural sciences, 1, 4, 16, 19, 111, 201
Netherlands, 205
Networks, 13, 28-30, 83. *See also* Fidonet; Internet II; Real-time
　monitoring and controlling, 4, 29, 34-35
　vulnerability of, 36, 39, 140-142
NGOs. *See* Nongovernmental organizations
Niche markets, 139
Nomenclature, 11, 83-88. *See also* Taxonomic definitions; Terminology
Noncopyrightable databases, 146-147
Nongovernmental organizations (NGOs), 81, 94-96. *See also individual organizations*

Nonvoluntary licensing. *See* Compulsory license
Nordic catalogue rule, 146, 179
Norway, 46, 183
Nuclear Database (NUDAT), 206
Nuclear Information Service, 118

O

Observational environmental sciences, 69-83
Observational sciences, 5-6, 50, 53, 201
Observations, simultaneous, 53
Obsolescence. *See* Electronic storage media
Oceanographic Society of America, 64, 216
Office of Management and Budget (OMB), 126, 131, 159
Office of Science and Technology Policy, 11, 15, 90, 102, 171
Office of Technology Assessment (OTA), 176
On-line transmissions, 156
Open publication. *See* Publication, open
Organisation for Economic Co-operation and Development (OECD), 10, 20, 27, 101
Organization of American States (OAS), 93, 108

P

Paleoclimatology World Data Center, 119
Pan American Health Organization, 93
Paris Convention for the Protection of Industrial Property, 135, 145, 170, 178
Patent and copyright systems, 8. *See also* Copyright law; Subpatentable innovations
Patent and Trademark Office (PTO), 156, 160
Peer review, 11, 101
People's Republic of China, 72-73, 205
Personal computers (PCs), 35, 110
Photocopying, 132-134, 155
Physical sciences, 1, 4, 19, 51-52
"Plain old" telephone service (POTS), 32
Policy. *See* Public policy; Science policy
Population issues, 74
Portable Operating System Interface for Computer Environments (POSIX), 33
Post, telephone, and telegraph ministries (PTTs), 40
Predigital status quo, 135-139. *See also* Digital versus analog data

Price ceiling, 14, 126, 129
Price differentiation, 6-7, 124-126, 129, 167-168
Price discrimination, 14, 116, 124-125, 129
Price of data. *See* Data
Print media, 140. *See also* Publishers; Scientific journals
Privacy issues, 48, 103
Private investment, 9
  incentives for, 141
Private monopolies, 116, 142
Private sector, 93, 96, 116, 120
Privatizing data. *See* Data
Processing data. *See* Data
Product differentiation, 14, 116
Professional societies, 5, 11, 49
Profit, 132-133, 166
Property rights. *See also* Intangible property; Intellectual property
  exclusive (*See* Exclusive property rights)
  protectionist, 14
Proprietary period, 11, 79
Proprietary rights, 48, 151, 162. *See also* Data, nonproprietary
Protectionist strategy, 155-156
  possible consequences of, 165
Publication, open, 17. *See also* Electronic publication; Professional societies; Publishers; Scientific journals
Public domain, 8, 166
Public good, 7, 15, 164-170, 202
  defined, 112-113
Public health and safety, 6, 56
Public interest, 5-6, 9, 17-18, 161
  exceptions to, 152, 158
Public investment, 14, 115, 128-129, 133
Public monopolies, 116
Public policy, 1, 7, 62, 124. *See also* Science policy
  debate over, 22
Public rights versus private rights, 8, 139-145
Publicly funded research and data, 3, 7, 17, 110-131, 133
Publishers, 135, 142, 154, 166, 168
  commercial, 5, 52, 135
  database, 139, 166

Q

Quality control and assurance, 5, 11, 71, 101

R

Ramsey pricing, 124
Reciprocity clause, 155
Recommendations, 10-15, 22, 43-44, 81, 83, 100-103, 128-129, 171
Regulatory approaches, differing, 145
Remote sensing, 204
Reproduction and distribution. *See also* Photocopying
  cost of, 79, 91
Research. *See* Applied research; Basic research; Collaborative research; International cooperation; International research; Publicly funded research and data; Users of research data
Retention, long-term, 5. *See also* Archives; Long-term data sets; Retrospective data sets
Retrospective data sets, 78. *See also* Archives; Long-term data sets; Retention, long-term
Return on investment, 48, 168
Reverse-engineering, 147, 156
Ribonucleic acid (RNA), 73, 212
Rights. *See* Derivative work rights; Intellectual property rights; Property rights; Proprietary rights; Public rights versus private rights
Royalties, 144
Russia, 205, 206

S

Sabre Foundation, 94, 109
Satellites, 25, 41, 58, 95, 110
Science agencies. *See* Federal government science agencies
Science and Technology Network International (STN), 211
Science education, 9, 13, 15, 92-93
Science Group Thermodata Europe, 209
Scientific data. *See also* Data
  accessing (*See* International access to scientific data)
  defined, 16
  rapid growth of, 3, 17, 57-58
Scientific issues, 47-109
  defined, 47-49
  progress in, 3
Scientific journals, 5, 12, 48-49, 52

Scientific research. *See* Applied research; Basic research; Collaborative research; International research; Publicly funded research; Users of research data
Scientific understanding, 82
Scripps Institution of Oceanography, 118
Search capabilities, 89
Secondary users of data, 106, 203
  supporting (*See* Documentation of data sets)
Seismic arrays and stations, 215
Semiconductor Chip Protection Act of 1984, 155
Shoemaker-Levy Comet, 36, 54
SIMBAD (set of identifications, measurements, and bibliography for astronomical data), 71
Simple Message Transfer Protocol (SMTP), 33
Simple Network Management Protocol (SNMP), 33-34
Simulations, 52, 65, 67
Soil science, 85
Soil Taxonomy system, 86
Soils and Terrain Digital Database (SOTER), 85-87
Sole-source data providers, 147
Space sciences, 53-54, 66, 71-72
Space Telescope European Coordinating Facility (ST-ECF), 196, 213-214
Space Telescope Science Institute (STScI), 213-214
Specialized Agencies of the United Nations, 10, 13, 44, 101
SPECINFO, 211
Standard Generalized Mark-up Language (SGML), 33
Standardization, 4, 11, 29, 33-34
Storage of data. *See* Data
Stratospheric ozone depletion, 55
Subpatentable innovations, 137
"Substantial investment," 151, 157
Sui generis regime, 137, 145-147, 149-160, 162, 164, 166-169, 203
Sweden, 205
Switzerland, 209

T

T-2. *See* Nuclear Information Service
Taiwan, 205
Taxonomic definitions, 11, 84. *See also* Nomenclature; Terminology

INDEX

Technical factors and trends, 1, 3, 24-40
  advances in, 4, 25-26
  concerns over, 35-40
Telecommunications. *See* Communications
Terminology, 83-88
TeX, 65, 88, 200, 208-209
Textual format, 20, 45
Thematic Mapper (TM), 122-123
Third World Academy of Sciences (TWAS), 96, 108
Trade-Related Aspects of Intellectual Property Rights. *See* Agreement on Trade-Related Aspects of Intellectual Property Rights
Trade secret law, 136
Trademark laws, 150
Transmission Control Protocol/Internet Protocol (TCP/IP), 33, 42
Transnational data flow. *See* Data, transnational exchange of
Tropical Ocean Global Atmosphere Program, 70
Turkey, 183

U

Unauthorized use, 158
Unfair competition laws, 139
Uniform resource locators (URLs), 66
  problems with, 23
Uniform resource name (URN), 23, 66
United Kingdom, 54, 66
United Nations Commission on Economic Development (UNCED), 44
United Nations Development Programme, 93
United Nations Educational, Scientific, and Cultural Organization (UNESCO), 86, 92-93, 108
United Nations Environment Programme (UNEP), 44, 85, 93, 99
United Nations Industrial Development Organization, 44, 96
U.S. Agency for International Development (USAID), 46, 93, 96, 107
U.S. Constitution, 8, 172
  enabling clause, 157
  First Amendment concerns, 139
U.S. Database Investment and Intellectual Property Antipiracy Act (H.R. 3531), 157-160, 185

U.S. Department of Agriculture, 93, 107
U.S. Department of Commerce (DOC), 119
U.S. Department of Defense (DOD), 31
U.S. Department of Energy (DOE), 73, 215
U.S. Department of State, 93
U.S. Department of the Interior (DOI), 118
U.S. federal government science agencies. *See* Federal government science agencies
U.S. Geological Survey (USGS), 57-58
U.S. government, 93. *See also individual government agencies*
U.S. House of Representatives, 8, 155-157
U.S. National Committee for CODATA (USNC-CODATA), x, 23
U.S. Supreme Court, 8, 154
Users of research data, 115-116, 125, 163, 169. *See also* Commercial users; Secondary users of data
  size of community, 14, 129

V

Value-added products, 7, 133, 140, 152, 164
Vietnam, 96
Volunteers in Technical Assistance (VITA), 94

W

Wireless communication, 4, 24, 28, 30
World Bank, 13, 44, 93
World Data Centers (WDCs), 10, 79, 94, 98, 101, 214-215-, 219
World Health Organization (WHO), 93
World Intellectual Property Organization (WIPO), 9, 15, 134, 142-143, 146, 155-156, 158, 170-171, 184
World Meteorological Organization (WMO), 55, 70, 93, 116, 216-219
World Trade Organization (WTO), 15, 155-156, 161, 170-171
World Weather Watch, 69-70, 216-218
World Wide Web (WWW), 26, 30, 42, 65, 68, 127, 204, 206

X

X.400, 33